MILITARY AIRCRAFT

Illustrations: Octavi Díez Cámara, Daimier-Benz Aerospace, Marta BAe Dynamics, British Aerospace, Eurofighter Jagdfluzeug Gmbh, Dassault Aviation, Saab-BAe Gripen AB, McDonnell Douglas Corporation, Military Industrial Group MAPO, Rosvoorouzhenie, Lockheed Martin-Boeling, Northrop Grumman Corporation, SRPC Zvezda-Strela, Hughes Aircraft, Texas Instruments Incorporated, Kongsber Gruppen ASA, Avibras Aeroespacial, US Navy, Pere Redón Trabal and Antonio Ros Pau.

Production: Ediciones Lema, S.L.
Editor: Josep M. Parramón Homs
Text: Octavi Díez
Coordination: Eduardo Hernández

I.S.B.N. 84-8463-108-7

Photocomposition and photomechanics: Novasis, S.A.
Barcelona (Spain)
Printed in Spain

MILITARY
AIRCRAFT

Designed as a lightweight and economical fighter, the development of a version for naval use demonstrated its qualities and possibilities as a multi-purpose airplane. During more than three million hours of flight it has made new records in safety, operational readiness, maintainability, and mission capability. This has created

notable sales success and continuous production which will be maintained until the end of the first decade of the 21st century.

Development

In the spring of 1974 the United States Navy made its VFAX program requirements known, for it to be equipped with a lightweight

OUTSTANDING QUALITIES

Its conception came about half-way through the 1970's, and with its flight qualities, associated systems capability, robustness, reliability as demonstrated to its operators, and its weaponry possibilities, the F-18 is an aircraft which few others can match.

TOTAL PRECISION

Involved in training activities or employed in operational tasks the «Hornet» demonstrates day to day its capability to carry out all kinds of missions. Here it is dropping free-fall bombs in the competition "Low Country Bombing Derby".

multi-mission low-cost fighter. It was obligated by a Congress decision to consider the prototypes YF-16 & YF-17 by General Dynamics and Northrop respectively, which were being evaluated by the Air Force.

While the comparative studies were being carried out, McDonnel Douglas proposed an optimized version, designated the F-17, in which Northrop was an associate partner. It was this which was chosen to be developed as a naval model, given the identification NACF (Naval Air Combat Fighter). The fighter version was renamed as the F-18 and the attack version as the A-18, but the conclusion was reached that the intermediate version F/A -18 could carry out both roles without any problems. It was therefore given the generic name F-18, and received the name Hornet.

Flight of the prototype

Specific development work began on the 22nd of January 1976, and on the 18th of November 1978 the first of eleven YF-18 prototypes flew. Two of these were twin-seaters with complete combat capability, but with 6% less fuel capacity in the internal tanks.

The first F-18A was delivered to the US Navy in May 1980. After evaluation trials on aircraft carriers, the first squadron was operational two years later. Included with these were the twin-seater type B versions, and with these «Black Knights» the Marine Corps reached full combat capacity. Navy pilots of the VFA-113 «Stinger» squa-

dron began to receive theirs on the 26th of March 1983. In 1985 part of the 14th Air Wing was incorporated aboard the aircraft carrier USS «Constellation» CV-64.

Advanced Versions

Between 1986 and 1987 the F-18C & D versions were produced, a combat single-seater and twin-seater respectively, the latter for a combination of uses such as combat or training. The «Night Attack» version was ready on the 6th of May 1988, it is compatable with the C/D type, including improvements to the avionics to allow precision attacks in any weather conditions during the day or night.

This version was the standard with few modifications until 1991, when a proposal for a bigger, more capable version was presented. This received the designation F-18E/F «Super Hornet». Work began in June 1992, and the prototype for this new version flew on the 29th of November

INCREASED POTENTIAL

The US Navy has the intemtion of purchasing between six and eight hundred F-18E/F <<Super Hornet>> aircraft, with which it will replace the older versions, increasing its operational capability in both air combat and precision attack missions, depending on what has been assigned to it.

OPERATING IN SPAIN

We can appreciate the lines of the Spanish F-18 twin seater which is shown flying above Zaragoza shortly after carrying out a training mission at the firing range at Bardena Reales.

MULTIPURPOSE COMBAT AIRCRAFT

The multipurpose possibilities of the «Hornet» for all kinds of combat missions, operating from both aircraft carriers and ground bases, have been the reason for its sales success with a large number of units purchased or with countries waiting to do so.

1995, carrying out the first aircraft operations with the aircraft carrier USS «John C. Stennis» CVN-74 on the 18th of January 1997. Production of these units began, determined by the budget, this same year.

Some outstanding characteristics

Manufactured by McDonnel Douglas, the F-18 is capable of playing the role of aerial defence fighter as well as a precision attack airplane, and its success lies in its multipurpose nature and operational capability.

The cockpit

The single and combat/training twin-seater versions both incorporate a cockpit which stands out above other lightweight fighters in active service at this moment, due to a rational and ergonomic design adapted to the requirements.

Three large CRT digital screens are noteworthy for presenting a combination of data in respect to radar, weapon systems, flight parameters and operation possibilities. At the same time there is a HUD system reflecting the platform's data so that the pilot doesn't have to stop monitoring

outside the aircraft. The pilot is sits on a Martin-Baken SJU-5/6 ejector seat, and is kept comfortable thanks to air conditioning and complete pressurization.

Propulsion plant

It incorporates two powerful General Electric double flow jet engines, with little interconnection. They contain some excellent general features such as, for example, the ability to engage maximum re-heat in only 3 or 4 seconds, good acceleration capability, and increased survival prospects in the case of engine failure.

The first versions, the F404-GE-400 model, with re-heat produces 7,258 kg of thrust. The most recent versions have the F-404-GE-402 EPE (Enhanced Performance Engine) which produces 7,983 kg of thrust, while the «Super Hornet», which has been evaluated with the F-414-GE-400, produces 9,979 kg of thrust. To guarantee its operation for the longest time possible existing aircraft incorporate internal fuel tanks with a capacity of 6,061 liters, together with three sub-wing fixing points to locate auxiliary tanks with a capacity of 1,250 or 1,810 liters. The F version has been designed to hold nearly two thousand liters of additional fuel.

Capability

It has a lot of potential for both aerial combat and many types of precision attack missions based on the acceleration capability supplied by the engines and its good combat qualities in various air-to-air and air-to-surface configurations. This is also taking into account the possibilities offered by its Hughes AN/APG-65 or 73 multi-mode radar, which detects targets located nearly a hundred kilometers away.

COMBAT MISSIONS

LOCATION	YEAR	PARTICIPATION	MISSIONS
• Libya/USA crisis	1986	Aircraft from from VFA-131, 132,314 & 323 squadrons.	-Aerial defence -Air-to-surface Guided weapon attacks of patrol boats & corvettes. -Interception of MIG 23 & 25, Mirage F-1 & Sukhoi aircraft.
• Kuwait invasion by Iraq	1990-91	F-18 aircraft of the Marine Corps, the Navy and Canada (Three F-18s were lost in combat).	Attacking ground & air defences -Shot down an Iraqi MIG-23 with AIM-7M "Sparrow" missiles from a F-18. -Sank various patrol boats by cannon fire or "Hornet" bombs. -Shot down two MIG-21s with AIM-9L missiles.
• Peace process in the former Yugoslavia	1994-95	Eight F-18A+ of the Spanish & United States Air Forces.	Aerial support operation for ground troops. -Use of: >BR-250 free fall bombs >GBV-16 laser guided bombs >AGM-656 "Maverik" missiles >AGM-88 "Harm" anti-radar missiles >Air to air missiles: AIM-9LI "Sidewinder" & AIM-7F "Sparrow"

In air-to-air missions the radar is able to locate aerial targets, and offers the pilot information as to which poses the greater threat and manages medium-range guided missile systems. Meanwhile, in air-to-ground missions, it supplies an accurate representation of the terrain over which the aircraft is flying and the targets which have to be hit and, in addition, allowing very low altitude tactical flying.

In addition the aircraft is also equipped with the powerful AN/AYK-14 digital computer, wich is the computer that identifys the threat. This made up of a display screen informing the pilot of the zone that the signals are coming from, guidance and tracking radars, and different decoy flare launchers. Using Tracor AN/ALE-40 for Spanish units, or Goodyear AN/ALE-39/47 for United States units; communication equipment in UHF, HF, DF; the Tacan Collins AN/ARN-118; the inertial navigator etc, electronic war equipment can be installed in pods below the fuselage if the mission demands it. These consist of laser designators, cameras for tracking and guidance, or tactical reconnaissance equipment.

The robust under carriage designed for use on aircraft carriers allows it to operate on unprepared runways. The arrester hook allows

NAVAL DEVELOPMENT

The «Hornet» was created as a naval airplane for which it incorporated basic characteristics for its use on aircraft carriers: a robust main undercarriage to support landings, the hook which allows it to grip the arrest cable and the two jet engines which guarantee maximum survival.

NAVAL FIGHTER BOMBER

As a naval attack machine the F-18 can launch its AGM-84 «Harpoon» anti-ship missiles against naval units up to a radius of 100 kilometers or employ the SLAM (Standoff Land Attack Missile) version to attack with precision every kind of surface target located within similar distances.

Weaponry

Depending on the version, the nine to eleven external fixing points allow the transportation of between 7 to 8 tons of arms. These include AIM-9 «Sidewinder» short-range air-to-air missiles; AIM-7 «Sparrow» or AIM-120 AMRAAM medium-range missiles, it is being able to carry up to six of the former and four of the latter. This is complemented by a multi-barrel Vican M61A1 cannon made up of six 20mm rotating barrels, fed by a loader of 570 rounds and with an continuous rate of fire of up to 4,000 rounds per minute.

Missions against surface targets allow the use of a wide range of arms, which among others include, the free fall bombs Mk82, Mk83 and Mk84; other bombs such as the GBU10 & 12, CBU-59 cluster bombs, roc-

it to catch the arrester cables when it lands on the deph of an aircraft carrier, the hidden nozzle located in the upper front part allows it to refuel in flight and other equipment consists of BITE (Built In Test Equipment), a self diagnosis system which considerably reduces maintenance work and localization of component failure in particular.

ket- launchers, and various types of AGM-65 «Maverik» missiles which allow precision attacks on targets; anti-radar AGM-88 «Harm» missiles, anti-ship AGM-84 «Harpoon» missiles with a range of some 100 kilometers and its multipurpose version SLAM; B57 tactical nuclear bombs.

EFFICIENT ENGINE

The propulsion plant of the C/D versions and earlier models have been modified, and consist of two powerful General Electric jet engines F404-GE-402 EPE (Enhanced Performance Engine) with low compression, reduced consumption and virtually no visible emissions. They are installed at an angle to line up the exhaust nozzles and are capable of giving a total thrust of 14,944 kilograms and of using both JP4 & JP5 as fuel.

ROBUST LANDING CARRIAGE

Designed for use on aircraft carriers, the front landing carriage has a new wheel and is very robust; in addition the two fuselage undercarriage units incorporate a system which allows the absorption of hard impacts. These features allow the "Hornet" to also operate from superficially prepared runways.

FOLDABLE WINGS

Designed for naval use, it is possible to fold the wings which facilitating transport operations and improving its positioning in hangers or maintenance workshops.

CHAFF LAUNCHERS

Survival during combat missions is guaranteed by its agility, the capability of its electronic equipment to neutralize threats and the use of chaff launchers incorporated in the underside of the fuselage. These fire cartridges which produce an electromagnetic or infra-red signature confusing seek missiles.

TECHNICAL CHARACTERISTICS

	F-18A	F-18C	F-18E
COST IN MILLIONS OF DOLARS:	30	38	40 approx.
DIMENSIONS:			
Length	17.07 m	17.07 m	18.31 m
Height	4. 66 m	4.66 m	4.88 m
Wingspan incl. guided missile-launchers	12.31 m	12.31 m	13.68 m
Width of foldable wings	8.38 m	8.38 m	—
Wing surface area	37.16 m²	37.16 m²	46.45 m/s
Flaps surface area	2.27 m²	2.27 m²	—
WEIGHTS:			
Empty	10,455 kg	10,810 kg	13,387 kg
Maximum	25,401 kg»	25,501 kg	29,937 kg
Max external load	7,711 kg	7,031 kg	8,051 kg
Internal fuel load	4.926 kg	4.926 kg	6,500 kg

External fuel load	3,053 kg	3,053 kg	3,053 kg
PROPULSION:			
Two General Electric engines	F-404-GE-400	F-404-GE-402	F-414-GE-400
Unitary power with post combustion	7,258 kg	7,983 kg	9,979 kg
PERFORMANCE:			
Ceiling service height	15,240 m	15,240 m	15,240 m
Speed at high altitude	+ Mach 1,8	+ Mach 1,8	+ Mach 1,9
Speed at low altitude	+ Mach 1	+ Mach 1	+ Mach 1,1
Runway length	427 m	427 m	380 m
Interception range	740 km	537 km	722 km
Extended range	3,326 km	3,326 km	4,000 km
Design load factor	7.5 g	7.5 g	8 g

OPTIMIZED COCKPIT

Designed to help the continual work of the pilot, it includes three large cathode ray tube display screens for different functions, A Head Up Display (HUD), a Martin Baker SJU-5/6 ejector seat, a flight control and weapons joystick and other complementary components.

INTEGRATED NOZZLE

On the front right side of the F-18 is the in flight refueling nozzle,automatically brought in & out, allowing it to receive fuel from tanker aircraft. Using this technique allows a significant increase in the radius of action for combat air patrol missions and those for COMMAO type actions.

ALL WEATHER CAPABILITY

Electronic equipment associated with the F-18 such as this FLIR (Forward Looking Infra Red) pod installed below the fuselage where medium range air-to-air missiles are normally carried, guarantees operational capability in any kind of atmospheric condition, day and night.

ADVANCED RADAR

The Hughes AN/APG-65 multi-mode digital radar is capable of tracking 10 aerial targets located within an 80 kilometers radius, presenting six of them to the pilot who can choose different air-to-air or air-to-surface modes depending on the mission which to be executed. It is a modular construction for maintenance purposes and is able to cope well with electronic counter measures.

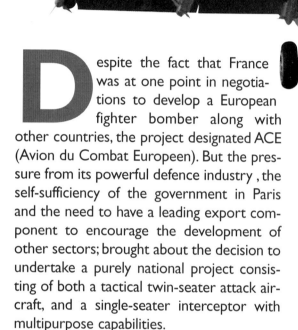

espite the fact that France was at one point in negotiations to develop a European fighter bomber along with other countries, the project designated ACE (Avion du Combat Europeen). But the pressure from its powerful defence industry , the self-sufficiency of the government in Paris and the need to have a leading export component to encourage the development of other sectors; brought about the decision to undertake a purely national project consisting of both a tactical twin-seater attack aircraft, and a single-seater interceptor with multipurpose capabilities.

The decision

While it was collaborating on the design concepts of the ACE, the Dassault company was continuing the development work on a new fighter bomber which could be its trump card for export. The first Rafale A

ON BOARD

In accordance with the requirements of the National French Navy, a specific model has been developed to operate on French aircraft carriers including the nuclear Charles de Gaulle, which is the destination for a squadron of Rafale M's which will constitute a multi-purpose component.

NAVAL VERSION

After evaluating the F-18 «Hornet» as an option, the French Navy decided to go for the naval version of the Rafale to make up the defensive-offensive component of its aircraft carriers. This was a process which required substantial changes to the components and equipment to cope with hard take-offs and landings and a hostile environment.

prototype flew on the 4th of July 1986, for which it had though it had to use general electric F404 engines, because the French engines would not being ready until February 1990.

A military aircraft

After the decision of the French military to incorporate the Rafale model in its airfields, the development contract for a single seater "C" prototype was signed on the 21st of April 1988, and shortly after a naval prototype was requested, redesignated M101.

The C model flew in October 1991 and the M version on the 12th of December the same year. It completed a high number of trials which validated the earlier expectations. These included both the launchings carried out by catapult in a United States base at Patuxent River and on the aircraft carrier «Foch». In February 1993 it was equipped with the new Thomson-CSF/ Dassault Electronique RBE2 electronic sweep radar (Radar a Bayalage Electronique) and the Spectra defensive system for the B model twin-seater.

Forecast

There is a market potential of 500 units, among which possible overseas orders can be found such as the existing proposal from the United Arab Emirates. The French Air Force & Navy have reduced their purchase expectations to 234 and 60 units respectively (with the first of these ordered on the 26th of March 1993). The M model aircraft

will be the first delivered with advanced equipment and avionics, ready for the nuclear aircraft carrier Charles de Gaulle, which being built at present and which is expected to be operational in the 14th fleet in 2002.

L'Armee de L'Air is not in a desperate hurry to have these aircraft, given that its current aerial capability allows it to carry out those missions entrusted to it, however it is expected that its aircraft will be delivered between 1998 and the year 2009. These will correspond to the multipurpose C version, and a training/low altitude penetration B version. This last version is foreseen as the aircraft to execute nuclear attack missions, provided with the advanced ASMP missile.

An operational aircraft

With three different versions developed from the beginning, all using the same fuselage, it is expected that this aircraft can achieve a useful life cycle of between 25 and 30 years; now that the program is being carried out following criteria with great potential and controlled development costs.

EVALUATION
The design and development of the Rafale has required the construction of different prototypes which combined different versions of single and twin-seaters, in accordance with the specific requirements of both the French Air Force and Navy.

SALES
The features of the Rafale and French sales policies box well for important sales success in those parts of the world where French military systems are already being used, although the high price for this aircraft will make its purchase difficult for countries with a modest budget.

Capabilities

With the capability of quickly changing its operational configuration, and able to operate day and night and in all weather conditions, its short and medium range missions include surface attack, aerial superiority, reconnaissance and high precision attacks using both conventional and nuclear weapons. For this the RBE2 radar is essential, having the capability to lock onto eight targets at the same time.

To carry out attacks the aircraft has an internal Giat DEFA 30mm cannon which operates at the rate of 2,500 rounds per minute, and it can hold more than nine tons of weapons on its 14 fixing points below the wings and fuselage (13 on the M model), these being made up of conventional and

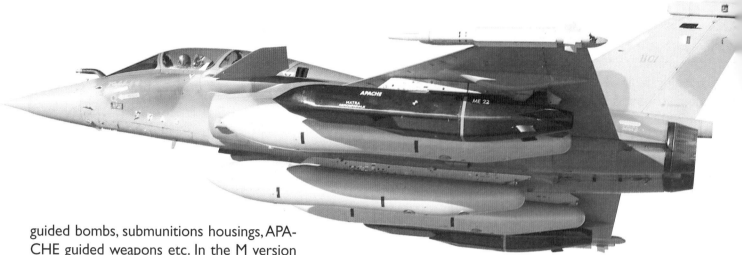

guided bombs, submunitions housings, APA-CHE guided weapons etc. In the M version up to eight medium-range MICA missiles can be carried supplied with infra-red or active laser guidance. For reconnaissance work it is provided with a pod equipped with electronic sensors and cameras, while precision attacks are executed with AS-30L guided missiles using laser designators with the Altis and supported by a FLIR system.

Integration of capabilities

The use of advanced technologies has meant the integration of different capabilities, constituting a truly multi-purpose fighter, small in size, which will replace more than six different types of specialist aircraft in France in the future. A system of processors have been integrated which facilitate the work of the pilot, and manage the greater part of the systems with the joystick (HOTAS concept, Hands On Throttle And Stick). It has been supplied with an integrated logistic support system (ILS), enjoys very efficient aerodynamics, incorporates various refinements to significantly reduce its radar signature and has equipment completely co-ordinated for self-protection, communication and management of weapons.

Advanced components

More than thirty specialist companies have been supporting the work to create advanced components, with the Rafale designed to incorporate all of these systems. The mission computers co-ordinate parameters related to the flight control system, the working of the engines, tactical

MULTIPURPOSE

With its design, equipment and performance, the Rafale fighter bomber marks the beginning of a new generation of aircraft designed to give high levels of performance in both air-to-air and air-to-surface missions and at the same time this airplane is capable of taking off from aircraft carriers.

INFRA-RED

The short range infra-red Mk2 «Magic 2» missile is a self defence component which can be fired from points on the outside edge of the wings.

systems, mission planning, self protection, vital support components, navigation, and coded data links, offer the pilot all the information necessary to carry out his mission and manage the weaponry.

A high technology holographic viewfinder with a large surface area, a large central display screen and two smaller complementary screens offer the user everything necessary to carry out the operational requirements, guaranteeing perfect integration between man and aircraft, thanks to the pilot's position. The OBOGS independent oxygen generating system; the high definition displays; the SEMMB ejector seats inclined at an angle of 29 degrees, with a viewfinder incorporated in the helmet.

Therefore, its structure, with two moveable stub wings located above the inlet nozzles, have been designed with maximum capability in mind. Among the materials employed the most striking are the com-

posites, titanium SPF-DB, superplastics, aluminium SPF, and different components resulting from the application of Stealth technology.

Control

The FCS digital flight control system ensures that the aircraft is stable at all times. Safe flying is guaranteed thanks to continuous and automatic control of flight modes and associated components such as the autopilot, altimeter, guidance system,

MEDIUM RANGE

Provided with their own infra-red or radar guidance systems, the MICA are the main aerial combat and interception weapon of the French Rafale.

CAPABLE

With 14 fixing points , its capability for carrying all kinds of weapons, bombs, housings, auxiliary fuel tanks etc, allows it to execute a wide range of missions, being easily re-configured from one mission to another.

etc. This capability is linked with the power produced by its two SNECMA M88 jet engines which despite being smaller and lighter than other conventional designs, are the result of an advanced research process which allow the aircraft to carry out long range missions without restrictions , with a particularly low level of fuel consumption. It has reached a high operational level and is of a modular construction which guarantees a low maintenance cost and also offers a high thrust-to-weight ratio.

MODEL C TECHNICAL CHARACTERISTICS

COST:	45 million dollars
DIMENSIONS:	
Length	15.3 m
Height	5.3 m
Wingspan	10.8 m
Wing surface area	45.7 m²
WEIGHTS:	
Empty	10,000 kg
Maximum	24,500 kg
Maximum external load	9,500 kg
Internal fuel load	4,500 kg
External fuel load	7,500 kg

PROPULSION:	
2 double flow SNECMA M-88-2 jet engines with a combined thrust of 13,800 kg.	
PERFORMANCE:	
Ceiling service height	18,000 m
High altitude speed	Mach 1.8
Low altitude speed	+ Mach 1
Runway length	450 m
Interception range	1,852 km
Extended range	4,000 km
Design load factor	+ 9 g's

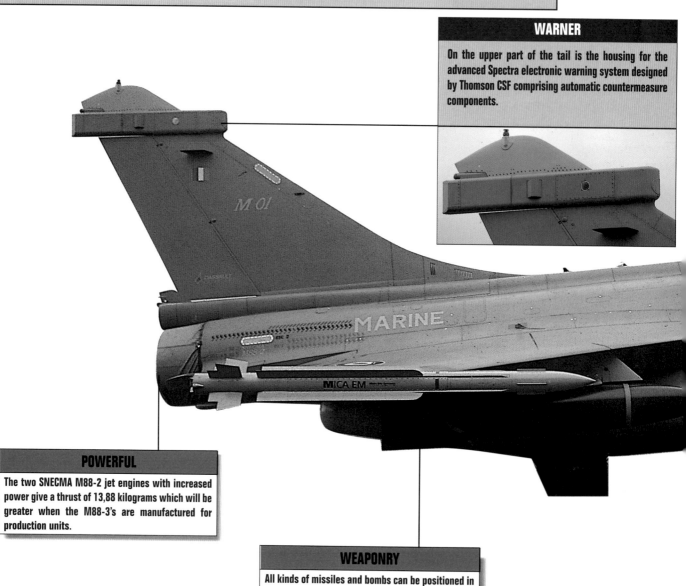

WARNER

On the upper part of the tail is the housing for the advanced Spectra electronic warning system designed by Thomson CSF comprising automatic countermeasure components.

POWERFUL

The two SNECMA M88-2 jet engines with increased power give a thrust of 13,88 kilograms which will be greater when the M88-3's are manufactured for production units.

WEAPONRY

All kinds of missiles and bombs can be positioned in the 14 fixing points to accomplish missions assigned to the aircraft, including housings for specific reconnaissance work.

LAUNCHABLE

For evaluation of the SEMMB Mk16 advanced ejector seat the front part of the fuselage has been used, moved at great speed along a track.

RANGE

With internal fuel tanks incorporated in the fuselage with a capacity of 5,235 liters, the auxiliary sub-wing tanks with 2,000 liters, the tank located under the fuselage with 1,700 liters and the in flight refueling nozzle, the aircraft has a tactical range which can be adapted as necessary.

EJECTION

Accomodated in a SEMMB Mk16 ejector seat, manufactured under license from Martin-Baker, inclined at an angle of 29 degrees, the pilot enjoys excellent visibility and the ability to be safely ejected at any altitude.

DETECTORS

A Thomson-CSF/Dassault Electronique RBE 2 radar (Radar a Bayalage Elecronique, deux plans) is housed in the nose of the aircraft, with the capability of following eight targets simultaneously, complemented by infra-red and optical sensors from Thomson-TRT/SAT OSF (Optronique Secteur Frontal).

LANDING CARRIAGE

Both the front under carriage with two wheels and the double rear units are very solid and resistant, allowing it to operate from a variety of runways including the small decks of aircraft carriers.

ADVANCED

The shape of the fuselage and in particular the two air intakes have been optimized to reduce as much as possible the radar signature, making it difficult to locate. In addition there is multiple on board equipment to improve its self defence capability against all kinds of threats.

The F-15 is considered to be the most advanced multipurpose fighter in service, boasting since entering service a record of one hundred combat victories without any aircraft lost. It caused the loss of 34 of 41 Iraqi aircraft which were shot down in combat during the Gulf war.

An advanced design

Knowledge of the Soviet Mig-25 interceptor's capabilities led the United States to program the construction of an aircraft capable of facing up to it, with studies for a new fighter with aerial superiority beginning in 1965. After evaluating the proposals of three different corporations, McDonnell Douglas was contracted to produce 18 single and 2 twin-seater aircraft for experimental trials. The first of the YF-15A single-seaters flew on the 27th of July 1972 and the first of the twin-seaters on the 7th of July the following year.

The contract

With the concept validated the first production units were contracted in accordance with the 1974 budget. The first of 572 A/B versions were received on the 14th of November 1974

and the last in 1979. This same year, on the 26th of February to be exact, the first in a series of close to 500 improved units flew identified by the letters C/D , holding an additional 907 kg of fuel.

After coming to a commercial agreement the Japanese company Mitsubishi was allowed to manufacture this model, and the first of two hundred F-A5J/DJs assembled in Japan were ready on the 26th of August 1981. These included radar warners and specially designed electronic war equipment. This was inferior to that of the United States, for which, from the 20th of June 1985 it was decided to apply the MSIP (Multi Staged Improvement Program), modernizing the Hughes AN/APG-63 radar to convert it to the AN/APG-70 model with greater memory capacity

EXCLUSION

With the capability of carrying more than eleven tons of weaponry, the F-15E can carry out exclusion missions where it will attack ground objectives over a long distance or play an air defence role against multiple targets, day and night or in bad weather.

IMPLEMENTATION

From its origins as an interceptor with a wide radius of action the «Eagle» has been evolved up to the point of being a very advanced multipurpose fighter bomber to which few countries can get access due to government and economic restrictions (photograph below).

MISSIONS

F-15C FIGHTER BOMBER	F-15E ATTACK AIRCRAFT
- Combat air patrols (CAP)	- Aerial superiority & defence
- Aerial superiority & defence	- Nuclear attack
- Long distance interception	- Bomb & missile attacks against surface targets
- Satellite attacks (ASAT)	- Defence against: cruise missile attacks (CMD) and ballistic missile attacks (TDM).
	Anti-aircraft defence suppression(SEAD) and reconnaissance (being evaluated)

and data processing, and installing a new Honeywell weapon control panel improving the defensive electronic war combination.

A multipurpose fighter

At the same time as the incorporation of new accessories to the aircraft in service, there emerged the need to evaluate the capability of the «Eagle», with respect to carrying out both attack missions and the gaining of aerial superiority, for which a twin-seater was modified. With satisfactory results from the trials, planners ordered the mass production of two hundred E version twin-seaters, the first of which was delivered on the 12th of April 1988 and based at Luke base in Arizona. The last was delivered during 1996.

Specializing in all weather precision attacks the «Strike Eagle», as it has been designated, operates with specific equipment which includes synthetic aperture radar, modified to give better resolution when it is working in a ground tracking mode, FLIR infra-red seeker, Martin Marieta LANTIRN attack and navigation system, made up of the AN/AAQ-13 navigation unit and AN/AAQ-14 attack unit. This at the

OPTIMIZATION

It's equipment, capability, and size give it multiple possibilities for deployment in both defence and attack missions in which, it should always be successful.

MULTIPURPOSE

Considered to be the most advanced fighter bomber at the moment, the F-15 is used by the United States, Israel, Saudi Arabia and Japan, countries which put a great deal of resources into maintaining armed forces with real capabilities for both deterrence and attack.

same time as subsequent cockpits including four display screens allowing the sharing of the workload among the two crew members, and the aircraft with a capability of carrying a payload of more than 11 tons.

Exportation of the F-15

The capability of this aircraft to carry out a dual role, as demonstrated during the Gulf War, led to the preparation of a proposal specifically aimed at export markets. Initially receiving the designation H. Still with, some limitations in the equipment and capabilities,

from 1994 it has been purchased by Israel with code F-15I, and by Saudi Arabia as the F-15S. The Japanese, however, decided to improve theirs with more advanced electronic equipment, faster and up rated processors, and with the 220E engine version which now offers greater power and substantial improvements to its performance.

To evaluate future concepts and to face up to the aggressive export policy of Russia, a modified F-15B was prepared, coming under the STOL/MTD (Short

Take Off and Landing/ Maneuvere Technology Demonstrator) program equipped with engines incorporating directional nozzles in two axes and advanced flow reversers, thanks to which it has demonstrated its capability to make ultra-short landings in spaces between 15 and 457 meters.

An advanced airplane

More than 1,200 A,B,C,D & E type units have been purchased by the United States Air Force, and those remaining in service, among front line units, the Reserve and National Air Guard , come to a total of 746. A larger number of these are needed in reserve, and are currently in AMARC storage zones at the Davis-Monthan base in Arizona. To these can be added the 90 A/B/C/D aircraft remaining in Israel from the hundred purchased, and the twenty five I aircraft optimized and in the process of being handed over. 147 were built in Japan under licence; 76 C and 72 S aircraft purchased by Saudi Arabia, 24 of which have the latest upgrades for aerial superiority missions, and 48 are prepared for surface attacks.

ATTACK

With its pay load capacity, range (due to the internal fuel tank load), and the potential of its systems to destroy any kind of target, this aircraft continues to be the standard- against which other designs are measured.

DEPLOYMENT

The USAF, Reserve & National Air Guard are often deployed to different parts of the world to participate in many tydes of exercises, demonstrating their potential to face any aggressor.

Its structure

With an aerodynamic design which allows it to reach high speed at high altitude, its Lear Astronics digital flight control enables it to carry out missions in an automatic terrain tracking mode. Structurally many of its components have been made with honey-combed panels of aluminium and graphite/epoxy fibres. For reinforcment there has been the extensive use of titanium and plastic derivatives; 60% of the E version's structure has been modified to allow for 16,000 hours of useful life at a high number of g's.

It has three undercarriage components, with one wheel for each and a pneumatic absorption system for operation on semi-

threat, this coming from a winning design and a combination of advanced systems. One of the more notable of these is the AN/APG-70 Hughes Aircraft Doppler Radar which operates in band X for the fighter version and L for the attack version, having a capability to detect and follow targets which are almost 150 kilometers away.

Its programmable signal processor detects and fixes on small flying targets at very high speeds at any altitude, and in close combat this target is hit automatically. The data from the sensors and information on the best weapons for attack it go through the McDonnell Douglas AN/AVQ-20 Head Up Display and are interrogated by the AN/APX-101 IFF Teledyne Electronics system to establish if the target

prepared runways. Two large tails, a large aileron on the upper center part of the craft, and two wings set back on the fuselage, all of which correspond to a design which is optimized for a response to current needs.

POWERFUL

The C & E version turbofans are very advanced, robust models providing agility in close aerial combat.

MODEL C TECHNICAL CHARACTERISTICS

COST:	55 million dollars		External fuel load	9,817 kg
DIMENSIONS:			**PROPULSION:**	Two F-100-PW-220 Pratt & Whitney turbofans with 10,770 kg of thrust each.
Length	19.43 m			
Height	5.63 m		**PERFORMANCE:**	
Wingspan	13.05 m		Service ceiling height	18,300 m
Surface Wing area	56.50 m²		Speed at high altitude	+Mach 2,5
			Speed at low altitude	Mach 1,21
WEIGHTS:			Runway length	274 m
Empty	12,973 kg		Interception range	1,200 km
Maximum	30,845 kg		Extended range	5,745 km
Max. external load	10,705 kg		Design load factor	9 g
Internal fuel load	6,103 kg			

Power

In this sense the two engines are located close together along the central axis , allowing the aircraft to continue to fly in the case that one engine fails. The F-100-PW-220 Pratt & Whitney jet engines have been substituted by the 229 model since 1991. This version offers 12,200 kilograms of thrust for each engine with re-heat. This gives it enough power to achieve a level of flight performance in which it acquits itself well in any combat phase. Additionally, it can take off from very short runways and climb almost vertically at a speed of 15,000 meters per minute.

Replying to threats

With more than enough power, it has an unequalled capability to face any aerial

TWIN SEATER

Multipurpose versions include a twin-seater cockpit, similar to the training unit, increasing mission possibilities with the pilot and systems operator sharing the workload.

is a friend or enemy. Surface attack data is presented on a multifunction screen and holographic display in front of the pilot, allowing the terrain to be followed accurately at low altitude, by using maps of the zone over which it is flying, which guarantes the successful use of weapons whether they are dropped or launched from the aircraft.

Weaponry

The pilots have an elevated position, sitting in zero-zero type ejector ACES II seats which can be launched at any altitude. They also enjoy excellent visibility both forwards and back. This capability allows them to execute their missions with a greater guarantees of success. The normal weapons carried for air-to-air actions include the General Electric M61A1 20mm cannon with 512 rounds, 4 AIM-9 «Sidewinder» infra-red missiles; 4 AIM-7 «Sparrow» and 8 «AMRAAM» AIM-120 radar-guided missiles.

In surface attacks the Dynamics Control Corporation AN/AWG-27 weapon control system is very useful and can be used to employ precision missiles such as the AIM-65 «Maverik», together with different bombs including the conventional Mk82 & Mk84; the Mk20 «Rockeye», the guided bombs GBU-10,12,15 & 25, the self-propelled AGM-130 for attacks at safe distances, and the nuclear B61.

ARMED

With a multi-barreled cannon incorporated in the upper part of the fuselage side and multiple fixing points below the wings and the central structure, the «Strike Eagle» can be provided with a wide range of weapon systems and associated equipment.

For self defence it relies on a Northrop Grumman AN/ALQ-135(V) automatic type electronic counter-measure system; Magnavox AN/ALQ-128 radar emissions detector; RWR Loral AN/ALR-56C equipment; Tracor AN/ALE-45 decoy flare launchers.

REQUIREMENTS

The size of the aircraft and its equipment possibilities require a large and specialized maintenance team at the landing base to guarantee its operational readiness.

The political, economic and military independance shown by Sweden in the last century and a capacity for self sufficency in matters of security. Have enabled it to develop an advanced and competitive industry which in the aeronautical area, it has been able to design the SAAB 35 «Draken» as well as different models of the Saab 37 «Viggen».

The necessity to face up to the challenges of the 21st century was the motivation for beginning this ambitious program, with the main objective of deseigning an advanced multi-purpose fighter which could complement the «Viggen» and enjoy the highest level of quality and capability.

MANOEUVRES

In flight the «Gripen» is capable of carrying out maneuveres at a very low speed as is shown in this shot where it is flying with two Mi-23-«HIND» helicopters.

ADVANCED

Designed for both air-to-air and air-to-surface missions, the features of the «Gripen» make it one of the most advanced of its kind.

This aircraft, designated JAS 39(«Jakt Attack Spanning», attack and reconnaissance fighter), is being produced and delivered to operational units, thus demonstrating a continued and advanced independence in the conception and construction of weapon systems.

The project

In June 1980 an ambitious project began with the conception, design, amd materialization of a lightweight combat aircraft which incorporated the latest advances in technology; and which was capable of operating in an independent way from its eventual base camps, such as specially prepared stretches of motorway or specific streets in urban areas.

In its design stage the name «Griffin» was chosen which refers to a mythological animal.

The requirements

After analyzing various proposals from outside the country, none of which matched up to the high operational requirements. The initial needs were put together in a document which was ready on the 3rd of June 1981. After obtaining government approval on the 6th of May in 1982 a development contract began with the JAS industrial group and FMV. The Swedish defence administration covered the construction of five prototypes and 30 production units.

With the program confirmed in the Spring of 1983, the evaluation of different components began with the radar and head up display, making use of «Viggen» type aircraft in the process. The initial results were very encouraging with all the time schedules being met including the delivery of a prototype on the 26th of April 1987, which then carried out its first flight on the 19th of December 1988. However, problems with the flight management system were detected, causing two accidents one of which meant the destruction of the prototype on the 2nd of February 1989, slowing down the trials.

Validation

External support was negotiated, principally from the United States, to solve the problems derived from inadequate software and others arising from the fuel system; the starting mechanism and avionics refrigera-

DEVELOPMENT

The «Gripen» Swedish JAS-39 fighter bomber has been developed to complement the specific requirements of the Swedish Air Force and has been offered to other possible clients, among which is South Africa, interested in its characteristics and potential.

LANDING CARRIAGE

The landing carriage of this lightweight fighter bomber was designed to operate from air bases as well as eventual airfields.

tion. The fitting out of the aircraft continued with, in 1989, the inclusion of a two seater version. The remaining four prototypes flew between May 1990 and October 1991.

Using two structures for the verification of fatigue trials, in 1993 a 16,000 hours research program was created to test its durability and lifespan. A short time before, on the 3rd of June 1992 to be exact, an order for 110 units of the second batch was approved which included improved software, a new flight control system, provision for the TARAS communication equipment, optimized processors and different camouflage.

Production

With the gradual delivery of the first production aircraft, with the first flight on the 10th of September 1992. The process of creating operational units began with the first 30 units being received at the end of 1996. The first twin-seater was also delivered earlier that same year. In June 1997 the Swedish government announced the purchase of 64 new JAS 39s from the third batch. These incorporated various updates including an improved engine, they were declared operational in September 1997. After three weeks of intensive exercises the F7 wing squadron was stationed at Satenas base.

If the number of units ordered doesn't increase and present production levels stay the same, 204 «Gripen» will be delivered to

12 operational squadrons before the year 2007; with the next squadron expected to have this aircraft being the Angelholm F10 wing. However, if we take into consideration earlier developments, it is possible that an additional batch will be incorporated into this number making it unnecessary to keep the «Viggen» in active service, which will improve the logistics chain and add mission capability.

A multi-purpose airplane

Conceived with specific mission criteria,

DESIGN
The fuselage, which incorporates large mobile fins to optimize its agility, has been designed for maximum power and reduced radar signature (photograph to the left).

ENGINE
The RM12 engine is a Swedish development of the United States F404 jet engines, but with a better performance (photograph to the right).

the aim being as much self-sufficiency as possible. The JAS offers many good qualities in the evolving world of aerial combat and with a notable capability as a medium range interceptor. It can be employed against surface vessels at sea and is capable of carrying out precision attacks in the support of its own ground troops as well as destroying important enemy targets.

Multinational

Although there has been the desire to maintain that the development of this airplane and the taking on of such an ambitious project is the fruit of capable Swedish industry, and the willpower of its politicians. The final cost would make such a decision unthinkable in the majority of European countries where such manufac-

turing is done together to reduce costs. The truth, not diminishing the efforts made by its own companies, is that they have had to turn to companies from the United States ,The United Kingdom, Germany and France who have supplied important components such as the propulsion system, flight control computer, flaps control, secondary power system, electrical generator, HOTAS control system, head up display, hydraulic system and cannon.

This was demonstrated half way through 1995,with the important agreement between Saab and British Aerospace, that a multinational collaboration reduces risks and avoids long and costly development. British Aerospace had participated in the design and manufacture of the wings from the beginning of the program, to produce a «Gripen» version for export to international markets. As a result they have presented various configurations, including different engines, to avoid possible United States sanctions to countries such as Chile, Brazil,

EXPORTATION

Being presented in recent international aviation show rooms, the Saab-British Aerospace JAS-39 «Gripen» has the possibility of getting export contracts from South America as well as the Middle East.

South Africa, Hungary, Poland, The United Arab Emirates and The Czech Republic. These could be fruitful with contracts consolidating the present production capacity.

Some advanced characteristics

Defined as the first fighter of the fourth generation put into service, this model presents straightforward operational qualities and simplified maintenance, allowing staff to carry out in 10 minutes all the operations necessary to allow a new mission to proceed. This involving the loading of new weapons, checking oil levels, the detection of possible system breakdowns, and checking of the main external components.

The PS-05/A multi-mode Doppler Radar is the fruit of collaboration between Ericsson Radio and the British company Ferranti, it relys on a carbon fiber arial which allow it to detect targets during ground, air and sea search missions, while following the terrain and avoiding obstacles it also controls weaponry such as cannons

TECHNICAL CHARACTERISTICS

COST:	Approx. 50 million dollars inc. development		External fuel	2.000 kg
DIMENSIONS:			**PROPULSION:**	**A RM12 WITH 8,165 KG OF THRUST**
Length	14.1 m		**PERFORMANCE:**	
Height	4.5 m		Ceiling service height	15,000 m
Wingspan	8.4 m		Low altitude speed	Mach 1.08
WEIGHTS:			Runway length	800 m
Empty	6,622 kg		Combat range	800 km
Maximum	13,000 kg		Extended range	3,000 km
Max. external load	3-4 tonnes		Design load factor	+ 9 g
Internal fuel	2,268 kg			

and missiles. Associated with this, and displayed on the three multi-functional CRT screens, is one for a FLIR infra-red sensor or laser, which facilitate reaching long range targets quicker.

A total of 40 processors have the task of monitoring the different flight and system parameters, including a triple digital flight system from Lear Astronics, HOTAS controls system, optical diffraction Head Up Display (HUD), a conventional joystick control, with artificial touch designed by the British Page Engineering. The ejector seat is a Martin Baker S10LS.

Mono-engine

Based on the United States General Electric F404, the Swedish companies Volvo Air and Saab have developed an optimized version designated RM12, with a length of 404 centimeters and a weight of 1,055 kg. Thanks to the greater volume of air and a higher operating temperature they have managed to increase the performance to 8,615 kg of thrust, being able to reach Mach 1.08 without the need to resort to re-heat.

With simple and economical maintenan-

WEAPONRY	
MISSION	**COMPONENT**
Air-to-Air	Short range «Sidewinder» infra-red missiles deignated 'L' or RB 74 forthe Swedish designation.
	Soon there will be the Iris T medium range AIM-120 «AMRAAM» guided missile
	In the near future is the medium range «Meteor» guided missile.
Air-to-surface	Conventional and laser guided bombs.
	Rockets
	Submunitions magazine
	AGM-65 or RB75 «Maverick» guided missiles
Anti-ship attack	Saab RB15F independent missiles
Various	Mauser BK27 de 27 mm cannon incorporated in the fuselage

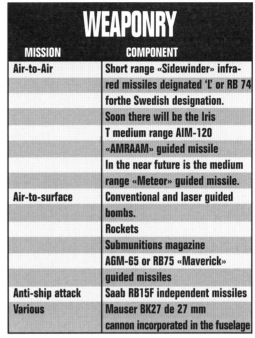

ARMED

All kinds of missiles, bombs and rockets can be installed in the «Gripen» enabling it to execute the most complex of combat missions.

COCKPIT

The cockpit has advanced digital displays and rationally configured equipment, with all functions optimized for the pilot's use.

ce, the new engine includes an electrical and mechanical fuel system with a consumption of 4.2 kgs of fuel per second when producing maximum power. It includes a larger diameter fan which is 15% lighter than the original the front end of the engine has also been made more robust as a protection measure against a bird strike.

Looking ahead to the future. They have begun making contacts with the German company Dasa and with the United States Boeing/Rockwell International consortium to input the necessary technology which would allow the development of a RB12 engine version. This has vectorized thrust TVC (Thrust Vector Control), initially destined for the export market.

onceived as an advanced multi-purpose fighter bomber. The Eurofighter will make up the main combat element of the air forces of Germany, Spain, Italy and The United Kingdom.

Design Philosophy

It has been constructed using advanced materials including, from the beginning, an Integrated Logistic Support (ILS) system for its 15,000 components. 70% of its fuselage is made of carbon fiber composite materials, 15% is metal some of which are aluminium-lithium alloys, 12% is re-inforced plastic and 3% is made up of other materials.

To comply with ESR-D requirements the single and twin-seater versions (the latter maintaining full combat capability) of the Eurofighter were conceived as a supersonic twin-engine airplane with delta and canard wing configurations which make it an extremely agile and capable machine, carrying out combat maneuveres which were not possible in earlier models. Special emphasis has been put on the design of a low weight wing increasing its relationship with the thrust; a minimized radar signature; an ergonomic,

advanced cockpit which gives the pilot the possibility of excellent external vision; equipped with a sophisticated collection of systems for attack, identification and defence.

Advanced sensors

Among these a third generation ECR 90 multimode digital radar has been developed which evolves from the radar used to equip the Tornado ADV (Air Defence Variant) and that installed in the Sea Harrier F/A2, This radar has been produced by the Euroradar consortium made up of the British firm GEC Marconi, the Spanish ENOSA of the INDRA group, the Italian FIAR and the German Telefunken Systemtechnik. It is a Doppler system incorporating important technological advances with its arial, transmitter and signal processor. Complementing these is an

DEVELOPMENT CHRONOLOGY

YEARS	
• 1976	Independent Program Group raises the need for a new European combat airplane.
• 1980	CONTACTS ← France / Great Britain / German Fed Rep DIFFERENCES PROJECTS ← ATC-92 Rafale / EAP / OWN TKF-90
• 1983	Incorporation of Italy and Spain
	Signing of OEST (Outline European Staff Target) Feasibility phase
• 1984	Signing of EST (Feasibility phase)
	Basic characteristics - 9.5 tons weight / 2 jet engines with a unit thrust of 85 kilonewtons each
• 1985	Withdrawal of France
	Signing of ESR (European Staff Requirements): definition phase
• 1986	Creation of industrial consortiums - Eurofighter (aeronautical companies) / - Eurojet (engine companies)
•1988	Signing of the development phase for the EFA prototype, jet engine and weapon systems.
•1992	Reorientation of the program due to political reasons.
•1994-1997	Flight of the prototypes - single-seater Germany / - DA2 Britain / - DA3 Italy / - DA7 Italy / - DA5 Germany / - Twin seater DA6 Spain / - Twin seater DA4 Britain: EJ 200 engine & ECR-90 radar

MASS PRODUCTION

The delay in the budget authorization for the production of the Eurofighter in Germany meant its date for entering service was set back. However the budget for 1998 included the economic parts necessary to undertake the process of mass production (photograph above).

AGILE AND CAPABLE

Wings with a large surface area; quadrangular air inlet nozzles; flat canards to increase agility and an advanced cockpit design are some of the Eurofighter's characteristics, a machine which can compete with its French and United States counterparts (photograph below).

infra-red tracking and guidance system installed in the area in front of the cockpit on the left side, and complex electronic warfare equipment incorporated in the machine's own structure, identified by the abbreviation DASS (Defensive Aids Sub-System) and with the capability to give a reply, either automatically or manually, to multiple threats including a complete ESM/ECM system, various radar warning detectors, chaff and interference flare launchers.

At the same time it incorporates an avionics system which it optimizes as an element

for aerial superiority and maintains additional capability for ground operations made up of seven functional sub-systems with a total of 24 computers linked together by data bases and optical fiber cables, all working together to supply the pilot with the maximum capability of managing tactical situations as well as flying. To improve this last aspect there is a Fly-By-Wire ACT (Active Control Technology) digital control system based on various computers which automatically control the aerodynamics against instability, giving it the highest levels of agility.

This aspect is possible thanks to the incorporation of two EJ 200 jet engines which have been designed especially for this airplane and which have an improved reliability performance, along with reduced fuel consumption. Each one of these produces 14,000 pounds of thrust without re-heat and 20,000 when this is employed, controlled by a digital system which optimizes its operation in whatever functional state.

Advanced cockpit

The cockpit incorporates a MK16A Martin Baker ejectable seat and facing the pilot there are three large color MHDD's (Multi-Function Head Down Displays), usually including data on the tactical situation, information on the state of the various sub-systems and maps of the terrain combined with the positions of aerial traffic. There is also a HUD (Head Up Display) visor with the responsibility of supplying basic flight data so that the pilot does not have to look inside the airplane for the information. This is complemented by the display system which is incorporated in the flight helmet using the HMS (Helmet Mounted Symbology) system, where the helmet incorporates a night vision facility and protection against various optical threats.

The VTAS (Voice Throttle And Stick) system allows the pilot to easily carry out complex missions in extremely intense situations. The joystick allows more than twenty functions to be carried out with respect to the control of sensors, weapons, the management of defence systems and flight handling, whereas the DVI (Direct Voice Input) system can be used to change presentation mode from HUD to MHDD, select targets and radio frequencies, all of these functional changes carried out by voice.

Weaponry Capacity

There are a total of thirteen fixing points (excluding the three for attaching additional fuel), with four on each of the wings and five on the fuselage, allowing the loading of different weapons up to a maximum of 6.5 tons. In air-to-air missions it can carry up to six radar-guided medium range missiles-AMRAAM (Advanced Medium Range Air-to-Air Missiles) positioned on the fuselage or in sub-wing mountings; six short range infra-red guided missiles-ASRAAM (Advanced Small Range Air to Air Missile) in sub-wing mountings; a 27mm Mauser cannon incorporated in the fuselage; it can employ older missiles such as Sidewinder, Sparrow, Aspide etc. For air-to-surface missions there are seven fixing points for locating free fall bombs, guided bombs, air-to-surface & anti-ship missiles, laser designators and various assemblies of associated equipment.

SPANISH PARTICIPATION

Spain has 13% of the work associated with the Eurofighter program and has constructed the first of the twin-seaters.

Its capability, reliability and multi-purpose nature make this a very advanced air-plane for which reason the air forces of Australia, The Arab Emirates, and Norway have expressed their interest, meaning in the medium term a large number of export possibilities are foreseen.

ADEQUATE PROPULSION

The increased propulsion power allows take off to be carried out on short stretches of runway and for loads to be carried over considerable distances, aspects which increase its operational capability and mission possibilities.

TECHNICAL CHARACTERISTICS

COST:	72 million dollars		**Internal fuel**	4,000 kilograms
DIMENSIONS:			**External fuel**	4,500 kilograms
Length	15.96 metres		**PROPULSION:**	
Height	5.28		2 Eurojet EJ 200 jet engines of 9,000kg thrust each.	
Wingspan including missile launchers	10.95m		**PERFORMANCE:**	
Wing surface area	50m²		Service ceiling height	15,000 metres
canards surface area	2.4m²		Speed at high altitude	Mach 2
WEIGHTS:			Speed at low altitude	Mach 1
Empty	9.999 kilograms		Runway take off length	700 metres
Maximum	21,000 kilograms		Interception range	600 kilometres
Maximum ext. load	6,500 kilograms		Extended range	3,000 kilometres
			Design factor	9g

INFRA-RED SENSOR

These models have an infra-red sensor located in the front left part of the cockpit, which improves the possibilities of its use in any weather conditions.

MULTIFUNCTIONAL RADAR

The ECR-90 multifunctional digital radar has been designed as a component with high operational readiness, with improved reliability and requires minimum maintenance. Its transmission power allows it to continue detecting aircraft at distances of hundreds of kilometres even when the atmosphere is being saturated by electronic counter-measure equipment, and it is capable of discerning the potential of distinct threats facing it.

LOWER AIR INTAKE

The fuselage has a quadrangular lower section due to the design of the large air intakes for the jet engines. This characteristic improves its functionality in some aspects of use and at the same time facilitates the positioning of medium-range missiles along its sides. The forward landing gear folds back between the entrance for the lower nozzles.

OPTIMIZED CANARDS

At both sides of the cockpit two large canards have been positioned, controlled by the on board digital computer to optimize the aircraft's capability to maneuvre at any flight altitude. A greater angle of attack, reduced turning radius, improved flight control with increased loads and greater lift are some of the parameters which define the canard's qualities.

OPTIMIZED COCKPIT

The cockpit incorporates a MK16A Martin Baker seat which guarantees ejection at any altitude; there are three large color MHDD's (Multi-Function Head Down Displays) in which all the parameters necessary for the mission can be visualized; a HUD visor (Head Up Display) which presents data to the pilot while he is visualizing the tactical situation and the VTAS (Voice Throttle And Stick), combining all the elements for total control of the aircraft.

TAIL RUDDER

The tail rudder, with its large cross sectional area, presents some advantages when carrying out missions needing combat agility and attack capability.

AUXILIARY INTAKE

Air is channeled to a dynamic refrigeration system for many of the electronic systems, which are kept inside the fuselage.

ADVANCED ELECTRONIC WAR

The housings located on the peripheral edges of the wings contain sophisticated electronic war systems (DASS).

PROPULSION UNITS

Reduced fuel consumption, increased operational capability, increased power as much with post-combustion as without it, and a thrust to weight ratio of 10 to 1. These are some of the characteristics which define the EJ 200 engines. Of modular construction it incorporates advanced components and digital control which optimizes its functioning at any altitude and mode of employment.

WEAPONRY

Close to six and a half tons of arms can be located on the five fixing points on the fuselage and on the four on the wings. On three of these it is possible to locate auxiliary fuel tanks which increase the range of movement as much as the combat range. Short-range missiles, medium-range missiles, free fall and guided bombs, air-to-surface missiles, sub-munitions magazines, stand-off weapons, equipment housings etc, can be carried to enable the carrying out of any mission assigned.

onceived for use on large United States aircraft carriers, the Grumman F-14 naval fighter is notable for a satisfactory combination of agility, acceleration, speed and range together with the capability for using a wide range of weapons in short, medium & long range air-to-air missions. All of which make it one of the best fighter planes at the present time. It is predicted that this model will continue in operational service until at least the year 2010.

It is famous for its intervention in aerial combat against Libyan Sukhoi aircraft in the Sirte Gulf, which were shot down with relative ease, and also for being the primary aircraft used in the «Top Gun» school, improving the ability of United States naval pilots. The «Tomcat» commands respect from its opponents for its capability and performance.

A highly developed model

After the failure in the development of a naval fighter craft TFX (Tactical Fighter Experimental), the United States Navy, towards the end of 1967, decided to build

LATEST MODEL

The F-140 D «Tomcat» is the most advanced of the United States naval-fighters, incorporating notable technological advances which will allow it to carry out interception and attack roles until the end of the first decade of the 21st century.

NAVAL INTERCEPTOR

Conceived with the idea of being used on large United States aircraft carriers the F-14 has demonstrated its potential for twenty five years and it is expected to be in active service until the end of the next decade.

an interceptor for the fleet, giving it the abbreviation VFX.

The decision

Once the qualities of the Grumman 303-60 were tested out and various modifications made, it was chosen as the winning aircraft and identified as the F-14 «Tomcat». Charged with the construction of 12 development models, the first prototype began its trials in Calverton on the 14th of December 1970.

The construction of 497 units was undertaken, the first F-14s of the series being delivered to the NAS Miramar VF-124

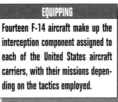

as «Bobcat» this improvement is based on the substantial modifications made to the weapons operator's display screen, and with the introduction of a variety of equipment including the LANTIRN (Low Altitude Navigation and Targeting Infra-Red, Night) and the IRST (Infra-red, Search and Track).

«Super Tomcat»

Between the 23rd of March 1990 and the 20th of July 1992 thirty seven new F-14D «Super Tomcats» were received with the capability of employing new missiles, night system compatible cockpits, advanced tactical display systems and the AN/APG-71 radar, having undertaken the modifications of 18 older units of this version. In parallel, modifications to other aircraft have been taking place to allow them to carry a reconnaissance pod called TARPS (Tactical Air Reconnaissance Pod) from a central hook up point. With this they are able to collect and send in real time digitalized images through a coded radio communication system, to any ship or ground position which uses the system Link 11.

In future it is expected that the fleet will incorporate the satellite positioning sys-

«Gunfighters» squadron. In 1973 a squadron boarded the U.S.S. «Enterprise» for operational assessment, some of these covered the evacuation of Saigon during the operation «Frequent Wind». The total number of units produced was 557, production ended in April 1987. A batch of 79 aircraft was destined for Iran of which twenty are maintained in operational condition, which were imposed, after the sanctions.

Modernization of the F-14

A few months before the end of production the first flight of the B version had taken place with improved avionics and propulsion, using the F110-GE-400 engine from General Electric. After initial trials the production of 38 new aircraft began. The first of which was received on the 11th of April 1988, and the last in May 1990, at the same time as the modification of some A types to a version originally designated as A plus, which was finally known as B.

This process continued until a short time ago, although Northrop Grunman has been involved in the supply of different equipment which allows the installation of additional capability to make precision attacks. Known

tem GPS (Global Positioning System), communication equipment from the AN/ARC-210 family, AN/ALE-50 interference systems, advanced mission recorders and digital flight control systems, among others.

A naval fighter

With a squadron of 14 «Tomcats» assigned to each United States aircraft carrier to supply the means of establishing combat air patrols (CAP) and making up the long range interception force, the F-14 is still considered to be a highly powerful fighter aircraft. 750 units have been manufactured of which 323 remain in service with others stored in a position of long term operational capability at Mountain Home air base in Arizona. A process which has required

INTERCEPTOR
With six AIM-54 « Phoenix» long range missiles located in the sub-wing and fuselage fixing points the F-14B carries out its missions in an interception zone with the objective of destroying any aerial target which comes within 200 kilometers of the established combat patrol area.

LANTIRN
Navigation and tracking equipment is carried in a pod fixed below the fuselage, allowing the destruction of surface targets in any weather conditions (photograph to the left).

POTENTIAL
With modernized avionics such as AN/ALQ-165 ASJP self protection equipment, the F-14D still constitutes a naval fighter of some note, with enough power to face up to later designs (photograph to the right).

external sealing and partial covering with a protective material.

An advanced design

The possibility of varying the angle of the external sections of the wings enhances its suitability for its role as an interceptor. The "arrow" can be adjusted between 20 and 68 degrees improving its aerodynamic efficiency and varying its lift. For naval use it has a robust undercarriage with a launch bar in the front wheel, a braking hook that catches a cable during landing and a nozzle which allows it to receive fuel during flight.

Located in the rear part just below the two large tail planes and the hydraulic air brake, we find the two F110-GE-400 General Electric engines which supply a total of 24,260 kgs of thrust, thus there is

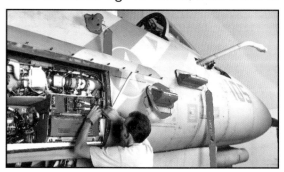

D MODEL CHARACTERISTICS

COST:	58 million dollars
DIMENSIONS:	
Length	19.1 m
Height	4.88 m
Wingspan with wings extended	19.54 m
Wingspan with wings retracted	11.65 m
Wing surface area	52.49 m²
WEIGHTS:	
Empty	18,951 kg
Maximum	33,724 kg
External max load	6,577 kg

Internal fuel	7,348 kg
External fuel	1,724 kg
PROPULSION:	TWO GENERAL ELECTRIC F110-GE TURBOFANS WITH 12,230 KG OF THRUST EACH.
PERFORMANCE:	
Service ceiling height	16,650 m
Speed at high altitude	Mach 1.88
Speed at low altitude	Mach 0.72
Interception range	800 km
Extended range	2,965 km
Design load factor	+7.5 g

more power, less fuel consumption, and better reliability than the originals, although they are still a little short on power.

Shared tasks

With an advanced capability for the time in which it was conceived, the cockpit combines various controls along the sides and analogue and digital gauges in front. Among the latter there is a Head Up Display HUD, a large display screen (Vertical Display Indicator) and a maneuveres control panel and in the pilot's hands a joystick which is moved in the conventional way.

For his part the co-pilot, who is given the designation RIO (Radio Intercept Officer), has at his disposition a large circular display (Tactical Situations Display) which allows him to see all the relevant information on the different radar screens, this offers a detailed picture of the tactical situation with the most serious threats being brought to his attention. There are various weapon control panels he can use; communication and IFF identification equipment; electronic war system; navigation data displays; fuel gauges.

Avionics

The oldest «Tomcats» rely on the powerful AN/AWG-9 Hughes radar which is capable of detecting targets at distances of up to 300 kilometers, to lock onto this data at 24 km and enter combat at 6 km. However, the latest units built have an AN/APG-71 monopulse with greater processing capacity which presents greater reliability and emission

ADVANCED

The capabilities of the «Tomcat» as a interceptor are due to the combination of advanced design and sophisticated equipment, which make it the most capable of its kind.

TWIN SEATER

The cockpit of the F-14 was designed for two operators with the responsibility of handling the aircraft as well as the weapon control systems, for which each position was designed with the equipment necessary for the task.

power. The power of the radar gives it the capability to be employed as a limited aerial warning alert system (AWACS), a task which the Iranian units still in service have been carrying out, despite the maintenance difficulties.

Along with this radar system there is the AN/ AWG-15F Fairchild Fire Control System; various data processing computers; the AN/ASW-27B digital coded communication link; the AN/AXX-1 television system TCS (Television Camera Set) and the IRST sensor located in the small housing inside the cone which covers the radar. AN/ ALE-29 Goodyear launchers- 39 for decoy flares; the AN/ALQ-100/126/165 ASPJ eletronic deception system.

The weaponry of the F-14

All aircraft carry a General Electric M6

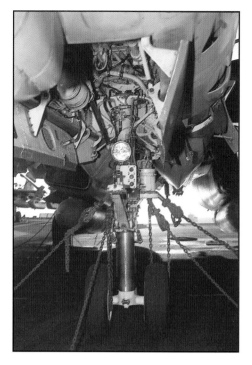

IAI 20mm cannon and a variable combination of missiles which can include four short range infra-red AIM-9 «Sidewinders», four me-dium-range radar guided AIM-7 «Sparrows» six long-range AIM-54C+ «Phoe-nixs». The latter is capable of intercepting targets with a 150 kilometer radius thanks to a design which combines its long range capability with an independent tracker which is able to distinguish between real echos and earlier ones coming from a variety of interference equipment.

Additionally some units, known as «Bobcat», have been modified to carry out attack missions in which different types of weapons can be employed including Rockeye & CBU-59 Cluster bombs, Gators & CBU-16 laser guided bombs, AGM-88 «Harm» anti-radar missiles with the possibility that they will be given the capability of launching «SLAM» missiles, a variant of the «Harpoon» anti-ship missile designed to reach armored targets located in enemy territory.

STRENGTHENED LANDING CARRIAGE

Take offs and landings on aircraft carriers demand that the landing gear is strengthened with the capability of coping with the enormous strain involved in this activity (photograph to the right).

PROPULSION

Despite the aircraft being of considerable weight the two General Electric F110-GE-400 turbofans transmit the power necessary to carry out attack and interception missions (photograph above).

DESIGN

Despite a design dating back to the 1970s the beauty of the Tomcat lines is undeniable, designed to give the best performance and speed in its basic role as a naval interceptor with the fleet, it is also an aircraft used in other kinds of missions.

nitially designed as an attack aircraft capable of counteracting ground threats resulting from a possible large scale attack by Warsaw Pact forces, the existing Tornado is a model which has evolved with various versions with great potential for the execution of air-to-air and air-to-surface tasks.

In this last activity it has been demons-trated as one of the most capable aircraft in existence at the moment being respon-sible for a great part of the low altitude missions during the first phases of the Desert Storm operation in the Gulf War.

GR4

With the capability of flying at very low altitude in any atmospheric con-dition, the GR4 improves penetration and attack capability compared with earlier models, with the result that the British Air Force has ordered the conversion of 142 units to this stan-dard. It is forecast that these will be delivered between 1998 and 2002.

MULTI-NATIONAL

Coming from a multi-national pro-gram, the Tornado is an excellent precision attack aircraft with a large part of the British IDS units being programmed for modernization to the improved GR4 version.

A multi-national aircraft

The development of the Tornado began half way through the 1960's in the core of the North Atlantic Treaty Organisation (NATO) with the idea of providing an air-craft capable of carrying precision attacks, which would fly at low altitude to avoid being detected by enemy radar and shot down by anti-aircraft systems. On this pre-mise, the 17th of July 1968 feasibility stu-dies began encompassing the governments of Italy, Germany, Great Britain, Belgium, Canada and the United States with, in the end, the first three countries continuing with the project. The development phase was completed in July 1970 and the struc-tural design in August 1972.

The first of the nine evaluation prototypes flew on the 14th of August 1974 and two years later, on the 29th of July, the governments involved in the aircraft pur-chase signed the production order for a total of 809 aircraft, these which would be divided into 6 batches, including the 6 pre-production aircraft. After establishing a common industrial manufacturing program in which a large number of companies participated, led by the British company British Aerospace, the

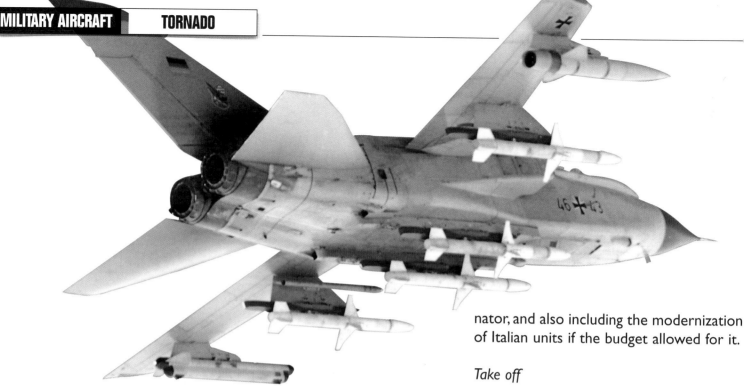

German DASA and Italian Alenia, decided to establish a multi-national training center at the British Royal Air Force base in Cottesmore, the first aircraft destined for Germany and Great Britain flew on the 27th of July 1979. The first for Italy flew on the 25th of September 1981 because it had delayed production approval.

Production

Supplies to operational squadrons began in 1982 and in March 1986 Britain received an order to deliver a total of 24 aircraft to Saudi Arabia. Three months later an order was signed for the production of a seventh batch consisting of 124 aircraft.

In 1989, coinciding with the decision to improve the British GR1 units, the latest IDS production aircraft for Germany and Italy were delivered. In January 1992, the last of the German ECR's were ready and in November this same year production for the RAF finished. In 1993 an agreement was made to transfer 24 British ADV aircraft to Italy. A second order from Saudi Arabia arrived in 1993 for 48 units which were due to be delivered in 1998, a year in which the MLI (Mid Life Upgrade) process for German units would begin. This included a new avionics structure and new equipment associated with the Rafael laser designator pod, or the infra-red FLIR desig-

---

RECONNAISSANCE

The ECR version developed by Germany is equipped with sophisticated optical and electronic equipment for reconnaissance work, allowing it find out the exact position of enemy threats which can then be counteracted by its own systems, through the launch of HARM anti-radar missiles or with disruptive emissions.

---

nator, and also including the modernization of Italian units if the budget allowed for it.

Take off

A total of 1,029 aircraft have been ordered including four presentation units some of which, since 1995, have been moved to the storage facility AMARC of the United States Air Force Davis-Monthan base in Arizona -as a result of the reduction of any threat.

The German Air Force operates 157 IDS aircraft, 55 training twin-seaters and 35 ECR's, shared out among squadrons stationed at no. 31 base at Norvenich, no.33 at Buchel, no.34 at Memmingen, no. 38 at Jever, no.51 at Schleswig/Jagel. Meanwhile the Navy incorporated 112 aircraft optimized for anti-ship attack, of which nearly half have

VERSIONS

MODEL	USER	IN SERVICE	MISSION
IDS	Germany, Italy, & Saudi Arabia	1982	All weather attack fighter bomber
IDS Navy	German Navy	1982	Anti-ship attack & zone reconnaissance
ECR	Germany & Italy	1990	Electronic warfare & reconnaissance
GR.1	Great Britain	1982	All weather attack fighter bomber
GR.1A	Great Britain	1987	All weather tactical reconnaissance
GR.4	Great Britain	1998	All weather attack fighter bomber optimized for night use
GR.4A	Great Britain	1999-2000	All weather tactical reconnaissance with capability for real time data transfer
ADV Mk3	Great Britain & Saudi Arabia	1986	All weather interceptor, aerial superiority fighter and combat patrol aircraft.

been transferred to the air force, grouped at Eggebeck Marinefliegergeshwader 2.

Italy obtained 88 IDS units and 12 twin-contro aircraft, 15 of which have been updated to the ECR version which allows them to launch AGM-88B Harm anti-radar missiles which were distributed among the 6th Stormo at the Brescia-Ghedi & Pratica de Mare bases. The 50th Stormo in Piacenza, and the 36th Stormo in Gioia del Colle, where anti-ship missions are carried out

PRECISION

The capability of flying at low altitude allows it to carry and launch Storm Shadow Cruise Missiles optimized for attacking high value, fixed targets whose position has been known some time before.

with Kormoran missiles. The 24 ADV Mk3 units on a ten year lease are shared among the Gioia del Colle and the 37th Stormo in Trapani /Birgi.

Great Britain purchased 164 IDS aircraft, 14 for reconnaissance, 51 twin-control units, 173 ADV (24 transferred to Italy), which it shared out to, among others, the squadrons at the Bruggen base, Coningsby, Leeming and Leuchars with, in 1992, a detachment of four aircraft going to the Falkland Islands. For its part Saudi Arabia is in the process of incorporating 24 IDS units, 6 reconnaisance, 14 twin-seater twin-control trainers and 48 ADV aircraft which are being shared among the 7th and 66th Dahram squadrons.

ATTACK

With an impressive capacity for carrying and launching all kinds of weapons against ground and naval targets, the Tornado IDS showed its possibilities during the Gulf War and has been employed in some missions above Bosnia. Missions in which its design concepts have been validated (photograph above).

POWERFUL

Provided with two Turbo Union RB199-34R Mk103 jet engines which give a total thrust of 13,500 kg, this attack aircraft can fly in an optimized manner both at low and high altitude (photograph to the right).

Specializing

Flying at 540 knots and at an altitude of 200 feet or less, British Tornado formations began the attacks against Iraqi air bases on the night of the 17th of January 1991.

Armed with JP233 submunition launchers its mission was to destroy runways,

damage enemy aircraft and hit all kinds of vital systems making it impossible for the adversary to respond. This was the multi-national operation Desert Storm which punished Sadam Hussein.

With significant losses among the attackers due to the numbers of weapons and missile defensive systems, the Tornado IDS 's demonstrated that they were capable of carrying out the missions they were designed for with complete effectiveness, optimizing its possibilities by varying the wing angle as a result of the speed and flight altitude.

For attack and defence

The most advanced version of the IDS model is the British GR.4 which incorporates numerous equipment systems allowing it to demonstrate its combat effectiveness with employing radar which could reveal its presence. On the left side, just below the radar cone is the infra-red FLIR GEC TICM II system, and in the front seat the pilot is seated in a Martin-Baker Mk10, relying on a

VARIABLE WING ANGLE

With the wings fully swept back the Tornado improves its penetration capability at high speed, while with them fully extended the lift is increased.

holographic type Head Up Display (HUD) which allows images to be superimposed, as obtained by the FLIR. He has at his disposition modified avionics compatible with third generation night goggles, which now include a new multifunction display screen offering information on both digital flight maps and associated systems. He relies on the accuracy which new satellite positioning equipment (GPS) gives him for his navigation purposes.

The co-pilot, with the responsibility for managing the weaponry, has at his disposition two large data presentation screens and a cartographic visualization screen, allowing him to carry out his missions without problems. For his support he can rely on a GEC-Marconi TIALD laser designator with which he can launch guided bombs from a safe distance, and at the same time manage an electronic counter measure system activated by a Marconi Zeus radar warner.

The ADV version, in the process of being modernized, include subsystems such as the

CHARACTERISTICS

	IDS VERSION	ADV VERSION				
COST:	35 million dollars	42 million dollars	**Internal fuel load**	5,830 l	6,580 l
DIMENSIONS:			**External fuel load**	4,500 l	4,500 l
Length	16.70 m	18.1 m	**PROPULSION:**	Two jet engines RB199-34R Mk103 with a unit thrust of 6,750 kg.	Two jet engines RB199-34R Mk104 with a unit thrust of 6,940 kg.
Height	5.95 m	5.95 m			
Wingspan					
minimum	8.6 m	8.6 m	**PERFORMANCE:**		
Wingspan			**Ceiling service height**	15,000 m	15,000 m
maximum	13.91 m	13.91 m	**Speed at high altitude**	Mach 2.2	Mach 2.2
Wing surface area	26.6 m²	26.6 m²	**Speed at low altitude**	Mach 1.22	Mach 1.22
WEIGHTS:			**Runway length**	900 m	700 m
Empty	13,890 kg	14,500 kg	**Combat range**	1,390 km	1,390 km
Maximum	27,950 kg	27,986 kg	**Extended range**	3,890 km	3,890 km
Maximum external load	9,000 kg	8,500 kg	**Design load factor**	+ 7.5 g's	+ 7.5 g's

AWS, ADMS and SPILS which optimize its working as an aerial defence aircraft and has been covered with absorbent sheets to reduce its radar signature. Operating with a Marconi Foxhunter Doppler radar which includes an IFF interrogator allowing it to detect its targets at a distance of 185 km. The pilot has responsibility for flight management while the co-pilot manages the air-to-air weaponry as well as the counter measure, interference and chaff equipment.

Offensive capability

With an external load capacity of 8 tonnes, the IDS version combines two 27 mm Mauser cannons with specific weapons such as MW-1 and JP233 submunition launchers; Paveway laser guided bombs; BL755 submunition bombs; Maverik laser guided bombs; Kormoran missiles; rocket launchers; free fall bombs; trusting in two Sidewinder AIM-9 missiles for self defence.

The ADV includes only one 27 mm cannon, the Sidewinders and four medium

INTERCEPTOR

ADV Tornados have been optimized for the interception of all kinds of aerial targets against which it can use Sky Flash radar guided air-to-air missiles.

DETAILS

A radar housed in the front nose; optimized air intakes for low altitude flight; in flight refueling capability; twin-seater cockpit provided advanced displays. These are some of the details of the Tornado (photograph to the left).

range Sky Flash missiles which in the future will be substituted by six Hughes AIM-120 AMRAAM's. Finally, this leaves it to be pointed out that normally electronic counter measure housings are installed, along with interference flare dispensers, Harm anti-radiation missiles and auxiliary fuel tanks.

Power

The attack version is equipped with two Turbo-Union RB199-34R jet engines, which in the Mk3 model produces 6,750 kg of thrust each, optimised to guarantee a flight without problems at low altitude where its automatic terrain following equipment, combining radar and inertial navigation, give it a high level of performance.

The model designated for aerial defence includes two Mk4 version jet engines with 6,940 kg of thrust each, while the electronic warfare ECR includes the Mk5 RB199 version which produces 10% more thrust than the initial model.

TAKE OFF

We can observe an IDS Italian Tornado on the ground at the Guedi air base, taxiing before carrying out a multi-national mission where it makes up a basic element of the Italian defence system (photograph to the right).

Designed to carry out landings and take offs vertically, thanks to vectorization of the jet engines exhaust gases. The Harrier presents outstanding features with the possibility of operating in multiple locations from large aircraft carriers to small clearings in a thick forest.

The creation of a military airplane

The idea of manufacturing a combat aircraft with vertical take off was developed in an era in which advances in the aeronautical world were the result of great interest in achieving greater and greater speed and capacity. Following this trend the British company Hawker Siddeley, in 1957, began the design of a military aircraft with the capability to land and take off on short

> **NAVY**
> Created to meet the specific requirements of the British Royal Navy, the Sea Harrier, which is also in service in India, has some differences to its multi-purpose brother, making it more effective in fleet air defence missions and surface target attack.

> **SPAIN**
> The AV-8B Harriers of the Spanish Navy are incorporated with the 9th Aircraft Fleet with approval to upgrade them to the Plus version around the end of the century.

strips of land. For this it collaborated with the company Bristol Engine, which specialized in high performance engines.

The Kestrel

After analyzing in detail the possibilities of project 1127, which managed to achieve the desired configuration, the initial proposal was tested in static flight at the end of September 1960, attached to the ground with steel cables. The good results obtained pushed the development, being followed with interest by Britain, the United States and Germany, who formed a tripartite squadron with nine P.1127 Kestrels.

After validating the concept and chec-

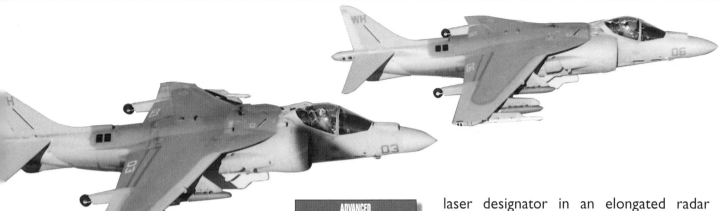

king its possibilities the British Royal Air Force (RAF) designated the preproduction units it received as the Harrier and those which followed for ground attack and reconnaissance were identified by GR.Mk.1, being delivered from 1967.

Development

After configuring the T Mk2 model as a training twin-seater version (T for training), optimization work began in aspects such as improved propulsion plants which took the place of the Mk1A & Mk2A which were a part of the Pegasus 102. In 1971 the first aircraft arrived for the United States Marine Corps (USMC), which designated them as AV-8A Alfa and TAV-8A, and then shortly afterwards the British Mk3 & Mk4 versions entered service which included a

ADVANCED
The Plus version of the AV-8B is capable of facing the most varied of threats, an aspect which has had a bearing on the fact that more than five hundred aircraft have been produced- to be complemented by the 82 type B's modernized to the new version in Spain and the United States.

MARINES
The United States Marine Corps relies on the capability of its Harrier B & B+ to supply the necessary support during amphibious assault missions.

laser designator in an elongated radar dome and radar threat passive detector in the front upper part of the tail. Later it was supplied with a Pegasus 103 jet engine.

Coinciding with the process of improving the aircraft, in November 1972 one of these flew to the Gulf of Vizcaya where it performed an exhibition for the Spanish Navy, landing and taking off from the then helicopter carrier Matadora. The result of this was the purchase of 12 single and twin-seater model A aircraft- the Matador, which were maintained in service until 1997, when they were re-exported to Thailand.

The Harrier on board

Just as the Spanish marines did, the United States and Britain decided to incorporate the Harrier on its specialized ships. As such, while the marines were using the V-8A, the Royal Navy, which had been experimenting with the P.1127 on the aircraft carrier Ark Royal during February 1963, went on to secure a

naval version- the Sea Harrier..designated FRSI to carry out missions as a fighter and attack-reconnaissance aircraft. The first units were delivered in June 1979.

After participating successfully in the Falklands War in 1982, in which the different Harrier models deployed were responsible for two thirds of the one hundred Argentinean aircraft shot down, it was made clear that more advanced models

THAILAND

Thailand has incorporated ten Harrier AV-8A single and twin seaters with the aircraft carrier Chakri Naurebet, these aircraft belonging to the Spanish Navy until half way through 1996.

were needed. Although at the same time the success brought about the contract with the Indian Navy, in January 1983, for the delivery of a total of 23 Mk.51 Sea Harriers, destined for the aircraft carriers Viraat and Vikrant.

An advanced aircraft

In 1984 deliveries of the McDonnell Douglas AV-8B Bravo and TAV-8B to the USMC began, with redesigned wings, greater weapon carrying capacity, more powerful engines. Advanced cockpit and a designator in the nose for launching precision bombs. 12 were incorporated by Spain in 1987 and adopted in the same year, as the Mk.5, by the RAF. In parallel to earlier deliveries the performance of the Night Attack prototype were evaluated. Adapted for night time combat thanks to the use of an infra-red sensor. The results were so good that the marines quickly decided to adopt it and the British began the transformation of their Mk.5 to this version, known as the Mk.7.

GR7

Manufactured by the British company British Aerospace, the GR7 is the most advanced version equipping the British Royal Air Force and is capable of carrying out attack missions both day and night. The pilot can be helped here by third generation night vision goggles or with the Infra-red FLIR tracker which supplies accurate images of the target.

With the operational possibilities coming to an end in 1990, the modernization of the British Sea Harrier to the F/A2 model was begun, incorporating a new GEC-Marconi Blue Vixen radar; with the fuselage lengthened by 35 cm; an integrated redesigned weaponry system and improved cockpit. Shortly after, the Americans, Spanish and Italians applied for the production and incorporation of the AV-8B Plus equipped with the Hughes AN/APG-65 radar; more powerful engines; decoy flare launchers in the upper and lower part of the fuselage; the cockpit modified with new display screens. The deliveries began in 1993, and also included a wide program of updating for the B series to the new standard in a process which will be finished at the beginning of the 21st century.

The Plus

Operating from air bases, eventual air fields, amphibious boats or small aircraft carriers, the Harrier has demonstrated its capability of carrying out multi-purpose

missions, with few differences between those units manufactured in the United States and those built in Great Britain.

The propulsion

The use of a Rolls Royce F402-RR-408 Pegasus 11-6 jet engine, producing 9,000 kg of thrust and including four engine flow deflectors located in pairs on both sides of the fuselage, permits these aircraft to perform in multiple scenarios. A rotary type which the pilot controls within normal limits, allows the escape gases to be directed, giving the possibility of vertical take off and landing, or the execution of agile maneuveres during close aerial combat missions.

Configuration

Designed for a useful life of 6,000 hours, the aircraft fuselage incorporate small differences with respect to other designs such as sturdy wings produced from carbon fiber composite materials, giving increased lift and greater capacity to carry fuel. Central fins replace the cannon housing when it is not being used. Large inlet nozzles improve the cruising speed and optimize the short lan-

---

PLUS

Designed to form a part of the Marine Corps and the Spanish & Italian Navies, the Harrier AV-8B+ presents greater potential than other earlier designs, relying on the multi-purpose Hughes AN/APG-65 radar, including an advanced cockpit, repowered engines and structural modifications to achieve better performance.

LOAD

With four fixing on each of the wings and fuselage, the GR7 can carry a wide range of weapons and support equipment for its combat missions.

---

ding and take off capabilities positioning of the support wheels in the wings which help the landing operation, with additional front and rear under carriages the inclusion of extended leading edges to improve the turning capability during air-to-air combat; the absence of rear nozzles, relying on the lateral deflectors.

Equipment

Sitting in a Martin Baker or UPC/Sentel ejector seat, the pilot has at his disposition an advanced cockpit which includes multifunction display screens and instrumentation appropriate to the missions.

It controls the air space and approaching land with an advanced multi-mode

WEAPONRY

AIR TO AIR MISSIONS	AIR TO SURFACE MISSIONS
A five barrel 25 mm GE GAU-12/U cannon or two 30 or 25 mm Royal Ordance Aden cannons. AIM-9L/M Sidewinder missiles and AIM-120 AMRAAM guides missiles.	Rocket launchers; general use bombs; BL 755 submunition bombs; Paveway laser guided bombs; multi-purpose AGM-65E Maverik missiles. The Hapoon anti-ship missile significant for its weight.

Doppler Hughes AN/APG-65 radar which is also the F-18 Hornet, and is able to launch its weapons accurately with the help of an infra-red illuminator; this model is equipped with a Smiths Industries HUD/HDD; instrumentation compatible with night vision goggles; a high precision inertial navigator; communication equipment resistant to electronic counter measures; a digital color map display; complete self defence equipment with interference cartridge launchers; advanced disruption system and a threat signaling display.

AV-8B PLUS TECHNICAL CHARACTERISTICS

COST:	50 million dollars	Internal fuel load	4,600 l
DIMENSIONS:		External fuel load	4,800 l
Length	14.55 m	PERFORMANCE:	
Height	3.55 m	High altitude speed	Mach 0.98
Wingspan	9.25 m	Low altitude speed	Mach 0.87
Wing surface area	21.37 m²	Combat range	259 km/h
WEIGHTS:			1,100 km/h
Empty	6,740 kg	Extended range	3,641 km
Maximum	14,061 kg	Design load factor	7 g's
Maximum external load	6,000 kg		

SUPPORTS

On each wing, supports are fitted for extra fuel tanks, and to carry the different types weapons necessary for executing missions.

HOUSINGS

The «Harrier A» is able to take advantage of auxiliary fuel tanks. Though in place of fuel tanks, containers carrying supplies for ground forces can also be fitted here.

AIRBRAKES

There is a large hydraulic air brake in the central part of the fuselage, this helps to reduce speed and makes the aircraft more maneuverable. A Tracor decoy flare launcher, firing infrared canisters to confuse the tracking missilesand canisters of multi-band CHAFF to disrupt enemy radar are also fitted.

FUSELAGE

The upper part of the fuselage incorporates various dynamic air intakes, which cool the internal electronic systems and engine exhaust nozzles. In addition there are markings and identification lights when flying in formation at right.

DIGITAL

The cockpits in the «Plus» variant, have incorporated notable improvements, including the management of the data presentation, with three multi-functionary digital screens, and a new Head Up Display (HUD). The improvements help the pilot to perform his mission more effectively, and make the likelihood of his mission being successful, a certainty.

REFUELING

Incorporated on the left side of the fuselage is the in-flight refueling nozzle which extends to receive fuel from a tanker aircraft, and is then retracted so that it doesn't effect the aerodynamics . With this process the mission range is significantly increased.

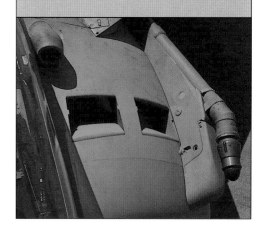

PEGASUS

Original in its design and configuration, the Harrier's Pegasus jet engine is distinguished from most other engines by its physical size, and by the size of the inlet duct. It is fitted with two swivel nozzle's which enable the pilot to direct the jet exhaust in almost any direction he requires. With this unique feature he is able to take off and land vertically.

LANDING CARRIAGE

Strongly built so that it can withstand heavy landing impacts, the undercarriage is fitted with a double wheel at the back to improve taxiing, It is also set back somewhat to allow it to be retracted into the fuselage.

The advanced lightweight single seater MiG-29 was the Soviet response to the need for providing itself with a modern aircraft which could substitute in a favorable way the earlier generation and which would have the qualities necessary to face up to the latest United States designs, while at the same time constituting an interesting aircraft for other countries. Its capabilities and possibilities have meant that it will maintain itself as a product with a key export role up to the year 2005. The last of these produced for Russia itself was in 1991.

An adapted fighter

Its operational and technical requirements were created in the LFI document

(Legkiy Frontovoy Istrebityel, a lightweight fighter for the front line) in 1972, with a view to replacing the MiG-21 &23 and the Sukhoi 15 &17, but it wasn't until 1974 that the order to build it came. The first of 13 prototypes flew from the Ramenskoye base on the 6th of October 1977 and was caught by United States spy satellites the following month, it was baptized in the West as RAM-L.

MiG-29 in service

A second aircraft was ready in June 1978, although like a quarter of the aircraft destined for evaluation it crashed due to engine failure. The production of the new model was begun in 1982 at the MAPO (Moscow Aircraft Production Organisation) factories. The following year delive-

ries to squadrons began, in 1985 it was considered to be operational and in July 1986 it could be seen in public for the first time when six of them flew at an exhibition in Finland. This allowed it to be examined up close for its flight agility and capability to be checked out.

The Fulcrum A, project 9.12, corresponds to the first version produced with three sub-types differentiated by small changes in some of the external details, while the B version, project 9.51, is a twin-seater trainer with total combat capability although it lacks radar. After its first flight on the 4th of May 1984, work began on the C type which has greater curvature where the fuselage joins with the cockpit, the aim being to carry more electronic equipment, possibly taken from other areas, improving the fuel load capacity by 75 liters, which in the original was significantly less.

Use with aircraft carriers

The K version identified as Korabelny began production in November 1989, which was to be flown exclusively from aircraft carriers. The extremities of the wings can be folded; it uses special RD-22K jet engines relies on a Zhuk radar. It has a greater fuel load and can carry various weapons some of the more notable ones being the Kripton AS-17 anti-ship missile.

Export possibilities

Advanced versions of the type C aircraft, with the designation S, have been produced since 1992 to improve export possibilities. With SD, SE & SM subversions this model includes digital flight controls;

DESIGN

The MiG-29's aerodynamics and mission capability have brought about great manufacturing success with exports to more than twenty countries, with Peru being the last to receive some.

AGILITY

The design, engine power and general mission capabilities allow this Russian fighter bomber to fight other Occidental designs in air-to-air missions from an advantageous position, enjoying a good reputation among its users.

missiles, including the SD's destined for Malaysia's MiG-29N. Adaptations for its use in tropical countries; satellite communication and navigation equipment; in-flight refueling system allowing it to receive 900 liters per minute; internal fuel tanks with 1,500 additional liters; structural improvements which allow it 28 degree attack angles compared with 22 degrees in the original; a RD-22 series 3 engine with a planned life-span of 2,000 hours.

Improved avionics

In the 1997 Salon Le Bourget a new configuration was presented with improved avionics, including four multi-function display screens- two large and two small for data

two multi-function display screens; a cockpit lengthened by 20 centimeters; eight sub-wing fixing points; and various modifications to the structural configuration for it to carry more fuel. A modernized Sapfir N019 (RP-29) radar with the capability of following two fighters at the same time (not just one as with the earlier versions); an updated Ts101M weapons computer; the possibility of employing more advanced

FIGHTER

The MiG-29 is armed with a wide range of short, medium and long range air-to-air missiles which together with it's internal cannon clearly demostrate the capability of Russian industry to design and manufacture advanced aircraft and weapon systems.

ters. Six hundred of these form a part of the Russian Air Force, with exports to some 2 countries to which Israel must be added which has purchased some for pilot training purposes, and the United States which ha recently ordered 21 units .

This outstanding sales success is due i part to Russian sales policies and the goo capability of the airplane, which can carr out many different kinds of missions and

presentation. Conventional instrumentation in a smaller size; a Topaz N019M radar provided with a terrain following mode and resolution of 15m; the possibility of employing AS-14 Kedge and AS-17 missiles, the first for surface attack and the second for anti-ship attack with active radar.

Sales success

Initially produced in the Znamya Truda and Nizhny Novgorod factories under the tutelage of MAPO, some 1,500 units have been built of which two hundred are UB twin-sea-

which requires less maintenance than other earlier designs. Despite having a small radius of action, relying on antiquated equipment and having been designed without considering the reduction of the radar signature, among its virtues are its maneuverability and agility in combat; excellent performance during slow flight; good visibility for the pilot who sits on a K-36DM ejector, inclined at an angle of 10 degrees helping him in the fast maneuveres which take place in close aerial combat and the acceleration of the two engines.

These engines consume a significant quantity of fuel and emit black fumes which are visible over a long distance. They are RD-33 jet engines produced by Klimov/ Sarkisov and are fed by large nozzles designed to guarantee the greatest flow of air at any speed, although their low position has necessitated the incorporation of a mechanical system, to close them when taxiing to avoid taking in foreign bodies which could cause damage.

In September 1996 a MiG-29 was displayed at the Farnborough Air Show in Britain supplied with RD-133 vectoral thrust engines, which thanks to nozzle movement in three directions gives it gre-

WEAPONRY	
GSh-301 Cannon	Very lightweight 30 mm caliber and with a capacity 150 shells.
KAB-500KR Bombs	Guided by television
BKF & ZAP-500 Bombs	Free fall. The first are high explosive and the second are Napalm incendiary bombs.
B-8M1 & S-24B launchers	These are for 80 and 240 mm rockets respectively.
AA-8 Aphid & AA-11 Archer	For use in air to air, short range and infra-red guided mission.s
AA-10A Alamo missiles	For use in air to air, medium range and radar guided missions.
AA-9 Amos missiles	For use in air to air, long range and radar guided missions.
AS-14 Kedge missiles	For use against surface targets.
AS-17 Krypton A y P missiles	Used against ships at sea and for anti-radar work respectively. It incorporates a search radar.

ter agility and increased potential for the uture.

n combat

During the Gulf War various Iraqi aircraft of this model confronted their western equivalents and suffered serious losses despite the pilots having red HMS displays incorporated in their helmets, these being very efficient and useful in close combat.

Although the cockpit is not ergonomic or comfortable, it can carry a wide range of weapons consisting of a cannon, different bombs & missiles supported by the capability of the OEPS-29 infra-red tracker which has a range of 15 km, which is integrated in the front of the cockpit; the Doppler impulse RLS RP-29 radar which has the capability to follow and attack targets flying below it, locating its targets within a radius of 100km and is equipped with a laser rangefinder. Despite these qualities, reports sent out by German pilots who have flown them mention the lack of power and the complicated handling.

COMBAT
Used by the Iraqis in Operation Desert Storm the MiG-29 could not show its qualities before the powerful Occidental war machine which was spread across the zone.

How it's equipped

With a Hydromash retractable under carriage and a body constructed from lightweight materials such as aluminium, carbon fiber, lithium and titanium, its equipment includes an integrated communication system with IFF detection; SRZ-15 interrogator and R-862 radio; INS navigation assistance; TACAN and GPS satellite positioning; self-protection system with a SPO-15LM radar threat warner; SO-69 transponders and a BVP-30-26M launcher for 26 decoy flares.

In addition the mission control system which has various computers linked with the radar, laser rangefinder, infra-red sensor and helmet designator, gives it capabilities which assure the destruction of every kind of aerial target at distances between 200 meters and 60 kilometers.

OPERATIONAL
The technology applied by Russia in aeronautical manufacture has substantially improved the operational level of its aircraft, which have demonstrated that they are capable of landing on foreign bases without excessive support or specific maintenance equipment.

VERSION C TECHNICAL CHARACTERISTICS

COST:	28 million dollars		External fuel load	2,200 kg
DIMENSIONS:			**PROPULSION:**	TWO KLIMOV/SARKISOV RD-33
Length	17.32 m			JET ENGINES WITH 8,290 KG OF THRUST EACH.
Height	4.73 m		**PERFORMANCE:**	
Wingspan	11.36 m		Ceiling service height	17,000 m
Wing surface area	38 m²		High altitude speed	Mach 2.3
WEIGHTS:			Low altitude speed	Mach 1.06
Empty	10,900 kg		Runway length	250 m
Maximum	18,500 kg		Interception range	600 km
Maximum external load	3,000 kg		Extended range	2,900 km
Internal fuel load	4,500 kg		Design load factor	9 g

PROPULSION UNITS

Located between the two vertical tails are the Klimov/Sarkisov RD-33 jet engines with 8,290 kg of thrust each, giving it great agility, although with a high fuel consumption it has a reduced range.

SUB WING

Sub-wing and fuselage fixing points allow it to carry a reduced payload, put at around two tonn. This is an aspect which has been substantially improved in the latest versions increasing this capacity up to four tons.

UNDER CARRIAGE

The under carriage is a solid construction allowing it to operate from semi-prepared grass runways, improving its capability to make landings in the most difficult combat conditions.

INFRA-RED

In front of the cockpit is an infra-red sensor which increases its detection capability, transmitting the data to the visor incorporated in the pilot's helmet, improving the designation of targets.

COCKPIT

Equipped with analogue and digital systems the MiG-29 cockpit is somewhat less advanced than its equivalents in the West, a point which the latest models substantially improve on.

RADAR

The latest Sapfir NO19 radars incorporate a tracking capability of two airborne targets with the possibility of also choosing the terrain tracking mode.

REFUELLING

To lessen the impact of the short-range of the first series of MiG-29's an in flight retractable refuelling nozzle was designed and incorporated to allow it to receive additional fuel from tanker aircraft.

AIR INTAKES

The large air intakes located at the front of the fuselage incorporate a mechanism to avoid the taking in of unwelcome objects during take off operations.

Following the tradition of earlier fighters, beginning with the Mirage III and F-I. The Mirage 2000 advanced fighter bomber was created as a more economical reply to the cancelled ACF project. It was designed to satisfy the needs of the L'Armee de L'Air and to guarantee the manufacturing independence of the French aeronautical industry. Widely spread throughout those countries which buy French military products, this fighter bomber has maintained a healthy manufacturing capability which makes France the third largest exporter of arms in the world.

Development of the Mirage 2000

Continuing with the proven design ideas of the delta wing Mirage III, V and 5/50, the designer Marcel Bloch, put his design philosophy into the development of this multi-purpose fighter, which gave it significantly more advanced qualities than earlier French models. Which although gaining outstanding sales success, showed themselves to be inferior to their comparable United States aircraft both in capability and reliability.

Development

After the cancellation of the ACF (Avion du Combat Futur) project for budgetary reasons, the development of a new combat aircraft to substitute it began in 1973. Since the mid 1980's the aircraft in

ADVANCED
The latest version of the Mirage 2000 to enter service was model 5 which incorporated an improved attack and navigation system combined with a new pilot aircraft interface, RDY doppler radar, integrated counter measure system, firing capability against various targets at the same time and the possibility of using a wide range of air-to-air and air-to-surface weapons.

the service of the French Air Force were manufactured by the company Dassault Aviation, centralizing the country's aircraft manufacturing capability.

The inaugural flight for the first of these took place at the Istres base on the 10th of March 1976. The fourth with a shorter tail was flown in May 1980, and the twin-seater in October of the same year. In parallel to these processes, there was from 1979 the production of a version specifically aimed at nuclear penetration, which flew on the 2nd of February 1983. On the 24th of October 1990 the multi-purpose 2000-5 was ready, being the latest model of this design,

USERS

MODEL	COUNTRIES (No of units)
2000B	Egypt (4), France (30), Greecce (4)
2000C	France (124)
2000N	France (75)
2000D	France (86)
2000DAD/DP	Abu Dhabi (6), Peru (2)
2000E	Abu Dhabi (22), Egypt (16), Greecce 36
2000H	India (42)
2000P	Peru (10)
2000RAD	Abu Dhabi (8)
2000TH	India (7)
2000-5	France (37), updated from earlier versions
2000-5E	Qatar (9), Taiwan (48)
2000-5D	Qatar (3), Taiwan (12)
Various	7 prototypes and 6 production units used by Dassault as demonstration aircraft.

incorporating noteworthy advances over earlier versions already in service.

Deliveries

22 aircraft were ordered as a result of the French defence budget in 1980, the first of which was destined for the l'Armee de l'Air and flew on the 20th of November 1982. The first operational squadron was formed on the 2nd of July 1984, when the EC 1/2 Cigognes from the Dijon-Longvic base reached the level required, with these aircraft being successfully used in the air-to-air and air-to-surface missions assigned to them.

Later, overseas orders arrived and the first operational squadron was announced in 1985. After this, orders for 550 aircraft arrived for all of the versions from the air forces of eight countries, half way through 1997 450 of these had been delivered.

CAPABLE

Designed to face multiple threats in an airspace saturated with electronic counter measures, the twin-seater D version incorporates a nozzle for in flight refueling and can be equipped with a laser illuminator to guide its 1,000 kg BGL Arcole laser bombs until they hit their target.

Going for a multi-purpose aircraft

After adapting the initial concept from being purely an interceptor to one with a more multi-purpose role to include ground attacks, the Mirage 2000 has demonstrated that it is capable of successfully carrying out multiple combat missions. These including long distance interception, aerial superiority, long-range penetrations, exclusion missions for enemy forces, attacking targets of great importance, the neutralization of air bases with anti-runway weapons, the control of naval vessels, reconnaissance, and tactical & strategic nuclear attack.

2000C TECHNICAL CHARACTERISTICS

COST:	34,5 million dollars		PROPULSION:	
DIMENSIONS:			A SNECMA M53-P2 jet engines with 9.690 kg of thrust.	
Length	14.16 m		**PERFORMANCE:**	
Height	5.2 m		Ceiling service height	16,460 m
Wingspan	9.13 m		High altitude speed	+ Mach 2.2
Wing surface area	41 m²		Low altitude speed	Mach 0.9
WEIGHTS:			Approach speed	259 km/h
Empty	7,500 kg		Runway length	457 m
Maximum	17,000 kg		Interceptor range	1,480 km
Max. external load	6,300 kg		Extended range	3,333 km
Internal fuel load	3,978 kg		Design load factor	+ 9 g
External fuel load	4,700 kg			

The capability of the Mirage 2000

The Mirage 2000 program has been able to maintain the French aeronautical industry as an international force conserving the important market share which it has traditionally occupied.

Equipment

The pilot sits on an Martin-Baker F-10Q ejector seat, manufactured in France under license. He has at his disposition a fly-by-

SERVICE

The Mirage 2000 makes up the main combat element of the French Air Force and has been purchased by seven other countries which trust in its ability to carry out every kind of mission.

ATTACK

The 2000N attack twin-seaters can carry out missions over long distances thanks to the possibility of positioning two large fuel tanks below the wings, allowing them to penetrate with weapons of every kind at low altitude above enemy territory and to fire from a safe position.

wire digital electronic flight control system, relying on the support of the SFENA auto-pilot with an advanced cockpit allowing him to travel comfortably thanks to the ABG-Semca air-conditioning system. In addition he has the support of equipment such as the VOR/ILS Socrat 8900, the SAGEM Uliss 52 inertial system, the TRT radio altimeter system, V/UHF & UHF communication systems and the IFF CNI NRAI-7A/11 transponder-interrogator, with many of these linked by a 2084 XR digital bus. The first models included a Thomson-CSF RDM multi-mode radar and the latest rely on the Dassault Electronique/Thomson-CSF RDI Doppler pulse

radar, both of which have an estimated range of 100 kilometers and with a high capability of working in atmospheres saturated with electronic counter measures. The French versions N & D incorporate a radar 5 version specializing in terrain tracking with a cartographic mode of operation, two inertial navigators and a GPS satellite positioner.

Propulsion

It incorporates a SNECMA M53-P2 jet engine which has been demonstrated to be solid and powerful during its years in service. This aircraft is capable of flying at a speed of around Mach 1 at low altitude, reaching more than Mach 2 when flying at high altitude and

POTENTIAL

The latest versions incorporate improvements such as more capable radars, in flight refueling nozzles, and an optimized cockpit all of which give it greater combat potential.

its acceleration capability is outstanding.

With 9,690 kg of thrust when using re-heat which is increased to 9,900 kg with the P20 version, and 6,500 kg when it is not used. It has a long operational range thanks to the fuel tank under the fuselage with a capacity of 2,498 liters and one under each wing with 1,480 liters. The twin seater has 74 liters less fuel.

Its weaponry

With nine fixing points, two on each wing and five on the fuselage, this aircraft can carry out both aerial defence and precision attack missions. The single-seater

APACHE
The capability of operating with the stealth air to surface APACHE missile gives the Mirage 2000 greatly increased potential allowing it to attack targets from a sufficiently safe distance, avoiding the reaction of anti-aircraft defences. In addition it has two Magic 2 infra-red missiles to defend itself in the case of another aircraft trying to intercept it.

EQUIPMENT
The upper part of the tail incorporates a Thomson CSF Serval receiver warning about other radars and in the lower part there is a housing for a parachute to facilitate breaking (photograph to the right).

WEAPONRY
The different fixing points below the wings and fuselage allow it to carry out all kinds of missions, with Magic 2 missiles used in air to air missions and AS-30 laser guided missiles for ground attacks (photograph to the left).

has two 30 mm DEFA 554 cannons (which the twin-seater does not have); two short range Matra 550 infra-red missiles; Magic/Magic2 or MICA IR on the extreme sub wing fixings; four Matra Super 530F or MICA medium range laser-guided missiles located in the fixings below the air intakes on the fuselage.

A multi-purpose fighter

As a multi-purpose fighter principally destined for air-to-air missions, it is equipped with an advanced Thomson-CSF RDY multi-mode radar which is simultaneously capable of detecting 24 targets and following the 8 which are more hostile. It relies on an ICMS integrated electronic counter measure system which doesn't require external housings and has a notably advanced cockpit. This includes a head up and head low display (HUD/HLD) in which the pilot can verify the tactical situation; two lateral data displays. A central display where radar supplied data can be presented and a flight control system based on the HOTAS principle according to which the pilot selects nearly all of the command functions without having to take his hand off the joystick.

Its air-to-surface weaponry combines 250 kg free-fall bombs; laser guided 1,000 kg bombs; Durandal & BAP 100 anti-runway bombs; Belouga submunitions dispensers; cruise & long range APACHE missiles; AS30L laser-guided & high precision missiles; Armat anti-radar missiles; anti-ship Exocets and in the nuclear penetration version N, the ASMP atomic warhead. The defensive system includes a Thomson-CSF Serval radar warner, an integrated counter measure system and a Alkan LL 5062 flare launcher.

COCKPIT

Created to facilitate the interaction between the pilot and the aircraft, the cockpit has been supplied with multiple multi-function display screens and a forward viewfinder which optimizes the pilot's work and the use of weapons.

The Mirage 2000-5

Recently put into production, the first aircraft of this model for France flew on the 26th of February 1996, Taiwan received its five units in April 1997, destined for the 2nd Tactical Fighter Wing. Qatar also gave an order, substituting them for its F-1's which Spain purchased. Doing so for a model combining up to date technological advances in its design, resulting in an advanced capability to face the threats of the next twenty years.

The Russian response to western technical advances in aviation has not been to sit idly by. Its latest aircraft are impresive in their design capability, maneuverability and sophistication. This widespread and well established line of new products has led the Russian firm Sukhoi to design models which can compete in an aggressive market. Its products, at the moment, are noteworthy with their considerable power, and it is hoped that they will enter the market in the near future, continuing the export tradition begun with the Su15, Su-17, Su-20 and Su-22 which have been sold to countries such as Afghanistan, Czechoslovakia, Algeria, Egypt, Poland, Syria, Yemen and Vietnam.

The Sukhoi's progress

Earlier aircraft became well-established leading the firm's designers to the development of a new attack aircraft model. Known at the beginning as the Su-19 «Fencer A», later in 1976, when the production of operational units began, this became the Su-24 Fencer B , with more than 900 units constructed in the Kom-somolsk factories. It has an outstanding

EXPORT

With the objective of obtaining the best external sales success possible, the Su-30K twin-seater has been built, existing in different multipurpose versions and exported to countries such as India and Indonesia, with a number of other countries interested in purchasin it.

capacity for neutralizing ground positions when flying at low altitude, with a flight profile optimized by varying the angle of the wings. Different versions have been developed, M with terrain tracking radar, MR with reconnaissance and electronic war equipment and MP with electronic disruption equipment, all of which were in service during the 1980's.

FLANKER

Optimized for air-to-air missions the Su-27 was the Soviet reply to western designs which entered service in the 1980's, and with its advanced configuration it demonstrates an agility which surpasses other fighter aircraft.

Lightweight attacker

Known for its initial role in the Afghanistan war, during which 23 were shot down, the Su-25 «Frogfoot» is a small airplane specializing in close support work and in the attack of troops on the battlefield. Operational since 1984 and exported to Afghanistan, Iraq,

Cze-choslovakia, Hungary and Bulgaria, it has served as the base for the UB operational conversion and weapons training units; the UGT for naval deployment; the BM equipped with aerial targets for fighter pilot training; and the UT which was pre-

sented in 1989 as the Su-28.

Developed from this last aircraft, the Su-25 was introduced in 1991, optimized for anti-tank missions thanks to its laser-guided missiles and 30mm double cannon. This along with the new Su-39 lightweight twin-engined attack aircraft, which with five fixing points for weapons and equipment on each of its wings and one on the underside of the fuselage, can carry 4.36 tons of weaponry which can be launched during day and night attacks due to the navigation and weapon firing systems incorporated.

Features of the interceptor

Initially known as the «Ram-J», after being detected by United States spy satellites, the Su-27 Flanker is a real indicator of Russia's design capability in high performance interceptors. After flying in 1981, more than two hundred units of this model

INTERCEPTOR

The design and components making up the construction of the Su-30K fighter give it some very good features for carrying out interception tasks it is capable of launching a large number of advanced weapons.

OPERATION

After 15 years in service with the Russian Air Force, a period of time in which it has demonstrated its operational possibilities, the design of the Su-27 has been optimized to serve as the base for a new generation of fighters such as Su-30, Su-35, Su-37.

have ended up in Russian combat squadrons and have been exported to China in various batches totalling some eighty aircraft. It is notable for its aerodyna-

mic shape and powerful engines which give it an agility which is superior to other occidental designs. At the same time it has a high maximum velocity, a large capacity to incorporate all kinds of advanced air-to-air missiles and a combat mission radius put at around 1,500 kilometers, all of which are excellent features for this long-range, all- weather interceptor.

Developed from the previous aircraft is the outstanding Su-30K fighter which is a twin-seater aircraft capable of operating in any weather conditions thanks to a Doppler type radar, which has the capability of detecting 10 targets in a 100km radius and engaging two of them it also has an optics and infra-red system for firing control; a maximum weight of 30,450 kg, of which 8,000 kg is the pay load and a maximum range of 3,000 km without refueling. In April 1997 the first 8 Su-30M's arrived in India, a multi-purpose version of the K model, with additional deliveries continuing up to the year 2001 with 32 additional Mk1 units. These include two AL-31FU jets with directional thrust control, in flight refueling capability, twelve positions for the loading of weapons and a maximum weight of 38 tons. Indonesia recently ordered 12 Su-30MK units.

The naval fighter

Designed along the lines of earlier models, the long range Su-32FN naval attack plane is a twin-seater with an armored cockpit arranged so that the pilot and wea-

pons operator sit side-by-side. The engines are significantly apart from each other to improve aerodynamics. This naval fighter has specific equipment and systems dedicated to it, such as a strengthened front under-carriage with two wheels allowing it to ope-rate from semi-prepared runways. The cock-pit incorporates a combination of analogue displays for flight parameters and four large digital display screens for tracking and attac-king.

Identified earlier as the Su-271B, it was exhibited at Bourget in 1995 and 1997, and is designated to attack naval targets such as hostile surface ships and submarines. Tasks for which it has an artificial intelligence sys-tem supporting the crew members in criti-cal situations. There is a specialized search radar and sonar buoy launcher with a capa-city of 72 units, magnetic anomalies detec-tor, infra-red equipment and laser rangefin-der, complemented with a wide range of weapons with which it can attack naval and aerial targets in a 250 km radius.

Looking to the future

The Russian Air Force is still deciding if it should incorporate the Su-35 into its aerial arsenal. This is an advanced version develo-ped from the «Flanker» with directional thrust and a multipurpose capability and whose inclusion is still not clear due to the government's financial problems. While the air force has been trying to make a decision, the new experimental Su-37 design has been flying since 1996, propelled by two AL-37FU Satum/Lyulka jet engines incorporating asymmetric nozzles which together produce 28,500kg of thrust. It has great agility, incorporating an electro-nic control system for all its moveable com-ponents, including the nozzles, as a result of the speed and altitude of the aircraft and with the N011M radar it can instantane-ously look for aerial and surface tarjets. Its basic characteristics are a length of 22.2 meters, weight of 25.7 tons and it carries the most advanced missiles among which

IMPRESSIVE
Large, heavy, robust, powerful and capable are some of the characteris-tics which define land-based naval attack plane Su-32FN being develo-ped by the Russian Air Force and which could pose a serious threat for occidental surface units.

OUTSTANDING
The capability of new designs coming from the Sukhoi laboratory has borne fruit with the display of a wide and varied range of models characterized by outstanding features among which can be found the Su-32FN designed as a long distance attack aircraft, replacing the Su-24 «Fencer».

are the air-to-air R-37 & KS-172, air-to-sur-face X-15P, and the stealth X-65S.

The existence of a new lightweight figh-ter became known as the result of the mock-up exhibited at the Paris Air Show in July 1997, a model which was similar to, but smaller than, the Su-35. The configuration includes canards, a twin tail and a centrally positioned Saturn AL-31F engine. It is equipped with a Sokol Falcon firing control radar with a maximum range of 180 km which will allow it to follow 4 of 24 targets it has detected. This aircraft could constitute the Russian response to the requirements of the 21st century with respect to agility, combat potential and performance.

Although it has to be developed for its qualities to be truly evaluated, some of the latest advances of this company are already known, including the fifth generation S-32/S-37, which in September 1997 first flew at the Zhukovsky Evaluation center where the noteworthy features were the inverted arrow wings, canards, and D-30F6 jet engines with vectorized thrust.

TAILS

Two large tails give it the stability necessary to execute low altitude flight above the sea, at the same time housing some secondary equipment.

PROPULSION

The two AL-31F jet engines which drive the Su-32FN produce a maximum of 25,000 kilograms of thrust, with which they can give it a maximum speed of 1.8 Mach.

ROBUST LANDING CARRIAGE

To satisfy the requirements of a long range naval attack airplane, it has been necessary to construct a very large and heavyweight aircraft which needs a robust main under carrige is retracted into the area where the wings and fuselage meet.

RADAR HOUSING

The large and elongated protuberance located between both engines can serve to improve general stability or for housing electronic mission support equipment, including the radar.

Su-32 FN TECHNICAL CHARACTERISTICS

COST:	Approx. 50 million dollars		PROPULSION:	TWO AL-31F JET ENGINES WHICH PRODUCE A UNIT THRUST OF 12,500 KG WITH RE-HEAT AND 7,500 KG WITHOUT.
DIMENSIONS:				
Length	23.3 m		PERFORMANCE:	
Height	6.5 m		Ceiling service height	18,000 m
Wingspan	14.7 m		Speed at high altitude	Mach 1.8
Wing surface area	62 m2	62 m²	Speed at low altitude	Mach 1.14
WEIGHTS:			Extended range	4,000 km
Maximum	44,360 kg		Design load factor	9 g's
Maximum external load	8,000 kg			

TWIN SEATER

With a duck-billed shape, the armored twin-seater cockpit houses the pilot and systems operator who together carry out the tasks entrusted to them with the support of display screens and control components, somewhat old fashioned compared with western standards but effectiv.

FUSELAGE

In the area to the rear of the cockpit is the lower entrance which gives the pilots access; the cannon which is incorporated in the fuselage; two large air intakes which feed the engines.

CANARDS

Behind and below the cockpit are two large canard fins which optimize flight at high and low altitude for this aircraft thanks to its movement which is controlled by the flight system.

Late in the afternoon of the 7th June 1981, eight airplanes left the Sinai from their Etzion base loaded with 907 kg Mk84 bombs. Flying close to the ground so as not to be picked up by radar, they crossed the 960 kilometers which separated them from Iraq's nuclear power station at Osirak, launching their bombs with precision onto the thick concrete walls and destroying the station completely. This combat mission, which was widely reported by the international press, was executed by Israeli F-16s.

Originally conceived as an aircraft which was easy to operate, economic and with some advanced features inherent in the original design, the General Dynamics F-16 has achieved notable sales success

MULTINATIONAL
Bought as a part of the Multinational program to equip four NATO countries, the Belgian F-16s make up the main offensive-defensive element of its airforce (photograph to the right).

SERVICE
Apart from forming part of the machinery of the Reserve and National Guard Airforces in addition to the air potential of a further sixteen countries, the F-16 has demonstrated that a well equipped lightweight fighter can carry out, without problem, air-to-air and air-to-surface missions (photograph below).

with some 4,000 units in service in nearly twenty countries.

A light weight fighter

Its development originated from the LWF (Light Weight Fighter) program of the United States Air Force, which advocated the construction of a simple daytime fighter of aerial superiority, light and economical which could complement the more

sophisticated F-15. In this way a prototype YF-16 was constructed which had its first flight on the 2nd of February 1974.

The selection

On the 13th of January 1975 the choice was announced and so began a process of engineering development on a grand scale which allowed it to meet all the initial requirements as well as other later ones which, proposed the installation of a radar, and the capability to attack surface objects.

8 pre-production units were contracted, 6 F-16A single-seater and 2 F-16B twin-seaters. The first of which flew on the 8th of December 1976, and the last on the 8th of June 1978 carrying out all of the evaluations at Edwards Air Force Base, California. In parallel Belgium, Denmark, the Netherlands, and Norway decided to adopt it and began manufacturing under license.

The production phases

Production of the A model began in the Spring of 1978 and the most important user has been the United States with 2,300 aircraft shared among the units of the USAF, the Reserve & National Guard with some hundreds stockpiled in a state of operational readiness over a long period of time. The first United States unit entered service in January 1979 in 338 TFW at the Hill Air Force Base in Utah, with production continuing in blocks of 1, 5, 19 & 15 until March 1985. Later they were modernised and incorporated the OCU (Operational Capabilities Upgrade) applied to the 15 block in 1987.

The C version entered production in 1984 in the 25 block and benefited from

the second phase of the MSIP (Multinational Stage Improvement Program) upgrades, which in parallel introduced substantial changes to the cockpit and aircraft structure. The third phase of MSIP was begun shortly after this. Deliveries of 30/32 blocks began in July 1986, these were different in that the first employed a General Electric F110-GE-100 jet engine and the second a Pratt & Whitney F100-PW-220. They were modernized between 1986 & 1992 with nearly three hundred units in the National Guard under the premises of the ADF program this gives them the capability to use medium-range guided missiles.

The delivery of the 40/42 block began in December 1988, and in 1990 three hundred were modified to carry out CAS/BAI missions. Improvements to the engines were carried out from 1991 to a variety of

different versions, and in May 1991 the MLU (Mid Life Update) was approved and applied to 300 units of the USAF and to aircraft of the NATO consortium up to 1999. In October 1991 the 50/52 block was delivered and from May 1993 these have been given the capability to launch «Harm» anti-radar missiles, going on to be called the 50D/52D. In 1993 the ground training GF-16A was accepted by the 82 Training Wing at Sheppard Air Force Base.

WEAPONRY	
AIR TO AIR MISSIONS	**AIR TO SURFACE MISSIONS**
- 20mm multi-barrelled cannon from General Electric, type M61A1 Vulcan, with a magazine of 511 rounds	
- short-range infra-red missiles:	- Free fall & laser guided
«Sidewinder» AIM-9 and «Magic»	bombs
- medium range guided missiles:	- CGPU 5/A
«Sparrow» AIM-7 & AIM-120	Housings with 30mm
	cannons milimetros
«AMRAAM»	- «Maverik» AGM-65
	missiles
	- «Harm» anti-radar missiles
	- «Harpoon» anti-ship
	missiles
	«Penguin» Mk3

Forty were modified and given the designation QRC (Quick Reaction Capability) and towards the end of 1995 went into operation in the skies above Bosnia.

An outstanding airplane

The notable sales success is due to the

combination of its advanced features and great potential in air-to-air or air-to-surface combat inherent in its original design.

The cockpit

The cockpit is notable for its design incorporating a Head Up Display (HUD) and multi-functionary screens, which present the data from the Westinghouse AN/APG-66 or the Northrop-Grumman AN/APG-68 (v) doppler radar which locates arial targets and locks the missiles on to it. In secondary missions this also includes tracking terrain.

The canopy is made out of one polycarbonate piece with a fine gold film applied this helps to dissipate radar transmissions, with the result that the aircrafts front radar signature is reduced by 40%, and providing the pilot an exceptional field of view. The McDonnel Douglas ACES II ejection seat is reclined at an angle of 30 degrees to improve the tolerance of the crew to increased gravity loadings- g's, which are

pulled during maneuveres. The joystick is located on the right side to facilitate the control of the aircraft while the pilot is paying attention to what is happening around him-ie the HOTAS concept (Hand On Throttle And Stick). This keeps the arm in a comfortable position while carrying out the mission. A four channel ditital flight control of the "fly-by-wire type" allows maneuveres to be executed with precision and speed.

Its design

With the wings incorporated into the fuselage, the aircraft has a robust and light-weight structure. This has allowed the fuel capacity to be increased while at the same time its support ability has been improved with increased angles of attack. The wing layout also incorporates two large flaps below the stern side of the fuselage, mounted at an oblique angle with the function of giving directional stability in extreme wind conditions.

With only one engine either the Pratt & Whitney F100-PW-220 or General Electric F110-GE-100. It is flown with great agility and fed through a single fixed-air intake

ANTI-SHIP

In accordance with the requirements of Norwegian specifications, the anti-ship missile «Penguin» Mk3 has been incorporated in the F-16, and with an operational range of 55 kilometers it can attack medium sized ships thanks to an explosive load of 140 kilograms. This gives an increased capability to this fighter bomber.

below the dividing plate under the fuselage. This positioning ensures the flow of fuel to the engine at elevated angles of attack, although it has the inconvenience of taking in foreign objects when operating from unprepared runways. Nevertheless it is utilized to house the robust three-piece landing gear, with the two main rear carriages going forward & up, and the front carriage pulled back and up.

TWINSEATER

With a similar capability to the single-seater the F-16B can carry out pilot training work in addition to combat missions.

C VERSION CHARACTERISTICS

COST:	22 million dollars		to 2.567 kg for
DIMENSIONS:			the D single seater
Length	15.03 m	External fuel	3.066 kg
Height	5.09 m	**PROPULSION:**	
Wingspan including		A Pratt & Whitney F100-PW-220 jet engine with 13,000 kilograms	
Missile launchers	9.45 m	thrust or a General Electric F100-GE-100 with 10,770 kg.	
Wing surface area	27.87 m²	An improved version with greater power has also been	
Flap surface area	2.91 m²	introduced , designated 100-PW-229 and F110-GE-129.	
WEIGHTS:		**PERFORMANCE:**	
Empty	8,273 kg with	Service ceiling height	15,240m
	F100-PW-220	High altitude speed	Mach 2
	engine and 8.627 kg	Low altitude speed	Mach 1
	with F100-GE-100 engine.	Runway length	360 m
Maximum 40/42 block	19,187 kg	Interceptor range	1,315 km
Maximum external load	5,443 kg	Extended range	3,890 km
Internal fuel	3,104 kg reduced	Design factor	9g

RADAR HOUSING

Housing the Westinghouse AN/APG-66 doppler radar dish or the more advanced Northrop-Grumman AN/APG-68, depending on the aircraft's year of manufacture giving complete capability for the tracking of various targets in air-to-air combat and missile guidance.

AIR INTAKE

Its design and positioning in a very low position allows the entry of the required airflow to guarantee the correct functioning of the engine at whatever speed and height, although there is the inconvenience of its position facilitating the intake of foreign objects during take-off.

LANDING CARRIAGE

Originally designed as a light weight and economic fighter, it was given a simple, but robust landing carriage. Its front undercarriage is located just under the large engine air intake with a small directional wheel which folds back and up , rotating 90 degrees in the process. The rear carriage is somewhat stronger.

COCKPIT

Provided with a polycarbonate canopy which gives the pilot good visibility, the cockpit has an angled ejectable seat improving resistance to the effects of gravitational pull and equipment which gives it the possibility to carry out multiple missions. A noteworthy part of this equipment is the Head Up Display (HUD) which appears in front of the pilot.

TAIL

UHF/IFF antennas are located on the front edge and on the upper part is the VHF antenna for high frequency communication. There is also a powerful rotating flashing light which acts as a beacon to avoid accidents.

NOZZLE

Below the tail, which includes electronic counter measure equipment and an auxiliary parachute for braking, there is a unique, high performance jet engine which delivers the power necessary for carrying out missions.

ALL WEATHER OPERATION

Housings with infra-red tracking equipment are attached to the underside of the fuselage allowing it to carry out air-to surface missions with precision day and night, increasing its utilization possibilities.

CAPABILITY

The rails at the wing edges, the sub-wing supports and the fuselage fixings all allow the transportation of a wide range of equipment including launchable weapons, fuel tanks, or housing to carry auxiliary equipment.

FMRAAM

The Hughes company, looking for the greatest destruction capability for all kinds of targets, has presented the FMRAAM, which combines all the components necessary to configure a medium-range missile with the capability of facing the threats of the 21st century.

D esigned to shoot down short, medium and long-range aircraft, they constitutes the main offensive components in air-to-air combat missions and are in a constant state of evolution to face threats which constantly change in character. They are used across a wide range of fighters and are optional in training, transport and specialist, aircraft, offering both a defensive and offensive capability.

Development

The need to shoot down other enemy aircraft during the first world war led to the machine gun being adopted as an offensive weapon. When this was demonstrated to be inadequate, it was substituted by many kinds of cannons. The performance of these and the changing nature of the threats led to various developments, such as the X-4 guided missile, which didn't reach production, developed by Dr Kramer in the Ruhrstahl factory at the end of the second world war. Then there were the 2.75" rockets used by the United States up to the middle of the 1950's which crystallized from the studies carried out by the United States Navy Bureau of Aeronautics to be supplied with a missile capable of shooting down airplanes flying at Mach 1.

NAVAL UNITS

The United States Navy equips its fighter bombers with a combination of missiles which include the medium-range «Sparrow» and the long-range « Phoenix».

The guided missile

Managed by McDonnell Douglas, the development took components from a naval anti-aircraft missile designed earlier and incorporated a radar guidance system which was finally evaluated in the Mugu Point Missile Validation Center in California. Produced since 1951 under the name Sparrow Model 1, its inadequate performance led to further alternative research programs by Raytheon and McDonnell, which ended up with the N-6 version of the missile. It entered service in 1958 in the F-38 «Demon» fighters.

In parallel to this development work engineers from the company Philco were working on the development of a missile which incorporated a five-inch non-guided rocket with a special head for hunting targets. After the first launches in 1953, the Sidewinder was ready three years later and in 1958 brought down its first aircraft in the Formosa crisis. A conflict in which a missile was fired against a Chinese aircraft which did not explode, remaining stuck in its fuselage. Shortly afterwards a Soviet copy appeared called the AA-2 «Atoll»

brought about after a detailed study of the captured United States weapon.

Weapon developments

The need to provide fighter aircraft with all kinds of particular missiles led to the beginning of the development of weapons such as the United States Bell Eagle, Hughes Falcon , Grumman Phoenix and the Soviet AA-1 Alkali, AA-3 Anab, AA-4 Awl, AA-5 Ash. The necessities brought about by the Vietnam conflict led to the introduction of much-improved and more capable missiles than before, at the same time that aircraft performance advanced in maximum speed, maneuverability, and the power of the on-board radars.

The design of a new generation of fighter bombers brought about the creation of missiles capable of carrying out quicker maneuveres and reaching their targets from

greater distances. Therefore the French created the Matra Magic and R530; the British the Red Top and Sky Flash; the Italian the Aspide, and the Israelis the Shafir, which have been kept hidden with new variants and more advanced models continuing up to the present day. At a time when aerial combat involves the use of a combination of airborne radar warning systems, communication, management and control in real

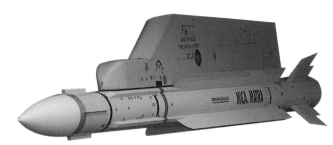

time, reconnaissance and detection satellites, and components reducing signatures based on Stealth technology etc, missiles continue evolving to face ever more complex and capable threats. Some of these are detailed in the following section.

Infra-red Advances

This relies on a sensor in the guidance head of the missile to lock onto the target, and includes a rocket engine with fixed or vectoral thrust to arrive at the target area and impact with it. It relies on fins and a very sleek fuselage to get the necessary flight sta-

tances in the majority of occidental fighters. With more than 150,000 units produced and different countries manufacturing it under licence, the Sidewinder has proven its effectiveness in the aerial conflicts such as the Falklands and Gulf wars with the more advanced L (or LI Improved), M, P, &X Evolved Sidewinder versions.

Similar to this is the French Magic, the latest being version 2 which has capabilities in every aspect with great maneuverability and the capability of being fired during any of the flight maneuveres of the host aircraft, giving it enough features to satisfy the widest range of combat necessities.

bility, and maneuverability allowing it to bring down the aircraft being attacked.

In the lower segment of this category are the United States FIM-92 «Stinger»AA, the Russian SA-18 Grouse, the British Starstreak and the French ATAM, a particular version of the Mistral. These were created for attack helicopters and airplanes in slow flight doing transport or reconnaissance

RUSSIAN
The Russian response to Occidental missile developments has not been to wait around, and now it offers a wide range of models which equal or exceed the capabilities of those offered by its opponents, and at a lower price.

A wide spectrum

Derived, in more or less the same size as earlier models, from those which in some

work, giving them a real air-to-air self-defence capability against similar aircraft. The mode of operation is the type "fire and forget" with the tracking head locked onto the target until it reaches it.

The «Sidewinder»

With a greater range and size than the others mentioned, multiple designs have been created following from the Hughes Sidewinder of the United States, it guarantees a kill over short distances with most western fighters defence over short dis-

cases have taken advantage of components or highly evolved characteristics we find the advanced Rafael Shafir 2, and Python 3- responsible for shooting down 50 Syrian aircraft in the Bekaa Valley. Also the Israeli Python 4, the Chinese CATIC PL-5, PL-7, PL-8, PL-9 which appear to be similar to the designs of other countries. There is also the Bra-

MULTI-PURPOSE
The Meteor and IRIS-T belong to a generation of missiles created to carry out the specific requirements of their producer countries, but also having good exportation prospects.

ZH590

The soviet surprise

The characteristics of the Vimpel AA-11 Archer missile caused a great stir in the West, equipping Russian designed aircraft, of which Germany has a group which made up a part of the MiG-29 arsenal of the old Democratic Republic. With a tracker capable of locking onto targets located at 45 degrees to its axis at a range of 30 km. With excellent maneuverability resulting from the vectoral thrust of its propulsion unit, it can hit targets when turning at 12 g's. It is much more advanced than was initially believed and sails past its western equivalents in terms of performance.

Future reply

To counteract it advanced designs such as the Matra Bae Dynamics ASRAAM (Advanced Short Range Air-to-Air Missile) have been worked on, which incorporates a tracker refrigerated by argon and nitrogen, is very resistant to counter measures, and has a long-range detection capability. There is also the German Bodenseewerk Geratetechnik IRIS-T with outstanding range and maneuverability and the French Matra MICA which combines a highly efficient combat missile head, an advanced

INFRA-RED MISSILES

DESIGNATION	ORIGIN	RANGE	MECHANISM	WEIGHT	EXPLOSIVE HEAD
MAA-1	Brazil	5 km	Active laser	90 kg	12 kg HE (*) fragmented
PL-9	China	5 km	Active laser	120 kg	10 kg HE
R-550 Magic 2	France	5 km	Radio frequency	90 kg	13 kg HE fragmented
MICA IR	France	50 km	Active radar	110 kg	12 kg HE fragmented
Python 4	Israel	15 km	Active laser	105 kg	11 kg HE fragmented
ASRAAM	G. Britain	15 km	Active laser	87 kg	HE fragmented
IRIS-T	Germany	12 km	Active laser	87 kg	11,4 kg HE fragmented
U-DARTER	S. Africa	8 km	Active laser	95 kg	17 kg HE fragmented
AIM-9X «Sidewinder»	USA	10 a 15 km	Combined	84 kg	10 kg HE fragmented
AA-8 «Aphid»	Russia	3-5 km	Radar or active laser	65 kg	6 kg HE fragmented
AA-10 «Alamo»	Russia	40 o 70 km	Active radar	254 or 350 kg	39 kg
AA-11 «Archer»	Russia	20 o 30 km	Active radar	105 or 110 kg	7,4 kg HE fragmented

* HE, High Explosive

zilian Sistemas Aeroespaciasis MAA-1 Piranha and the South African Kukry, Darter and U-Darter which will include vectoral thrust and a focal plane array tracker which distinguish between thermal images generated by the target and the background, being much more selective in the area chosen.

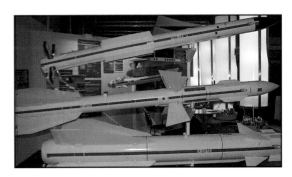

propulsion unit giving it a range of 50 km and, depending on the mission, the possibility of mounting an infra-red sensor or radar guidance unit.

Radar guidance

With a longer range than earlier versions, occasionally exceeding 100 km, capable of managing significant changes in altitude, a cruising speed greater than Mach 3, and a radar tracker incorporated in the guidance unit; these missiles are aimed at intercepting medium and long-range targets.

Made in the USA and spread throughout the world, the AIM-7 Sparrow being used at the moment corresponds to the versions F, M & P, provided with a monopulse, semi-

GUIDED MISSILES

DESIGNATION	ORIGIN	RANGE	MECHANISM	WEIGHT	EXPLOSIVE HEAD
PL-11	China	25 km	—	220 kg	HE fragmented
MICA AR	France	50 km	Active radar	110 kg	12 kg HE fragmented
Aspide	Italy	40 km	Active radar	220 kg	30 kg HE fragmented
AIM-7P «Sparrow»	USA	45 km	Active radar	230 kg	39 kg HE fragmented
AIM-120 «AMRAAM»	USA	50 km	Active radar	157 kg	22 kg HE direct fragmentation
AA-9 «Amos»	Russia	100 km	Active radar	490 kg	47 kg HE fragmented
AA-12 «Adder»	Russia	50 km	Active laser	175 kg	30 kg HE fragmented
R-37	Russia	150 km	—	600 kg	60 kg HE fragmented

active radar tracker, improved resistance to electronic counter measures and in the case of the M & P a fragmentation explosive head. Originating from this missile are the Italian Alenia IDRA/Aspide Mk2, the British BAE Sky Flash and various Chinese copies somewhat larger, designated PL-10 & PL-11 manufactured by CATIC, with the French counterpart being the Matra Super 530D.

The Hughes AIM-120 AMRAAM (Advanced Medium Range Air-to-Air Missile), guided by active radar, is the logical evolution from the Sparrow, with a range of 50 km and supplied with substantially improved electronics to face the threats forecast for the beginning of the 21st century. With an updated capability the Hughes AIM-54C+ is the most advanced version of the Phoenix which operates in conjunction with the F-14 Tomcat, guided by radar with a range of 150 km.

Faced with the varied number of weapons offered by Occidental countries, Russia has made great efforts to provide its

fighters, and those units for export, with a very wide range of missiles designed with a high level of performance. Among these are the particularly noteworthy AA-5 Acrid R-46, AA-7 Apex R-24R, AA-8 Aphid R-60, AA-9 Amos R-33 and its advanced R-37 version destined for the MiG-31 interceptor Foxhound, AA-1 Alamo R-27 in configurations which reach distances between 40 and 110 km, AA-12 Adder or R-77 AMRAAMSKI, and the AAM-L which can reach objectives located 400 km away.

SUPER «SIDEWINDER»

The Hughes AIM-9X is the United States industry's answer to the requirements of the US Navy for a short-range infra-red missile with an advanced tracking head, with control components which give it greater maneuverability, and a rocket engine with enough power to follow every kind of target.

OPERATION

With self protection infra-red missiles on the extremities of the wings which allow it to confront similar missiles, this Mirage F-1 is ready to begin a ground bombing mission.

Designers were led to incorporate a space in ship constructions to accommodate the transport, maintenance and operation of specialized helicopters. This came from the need to provide surface naval units such as frigates or destroyers with additional capacity in the search, location, and disabling of enemy ships. These can also be deployed from amphibious vessels, and aircraft carriers.

Assigned to special missions, naval helicopters carry out an important role, increasing the possibilities of action for the units they are deployed from. Giving greater capacity to carry out anti-submarine (ASW) and surface-to-surface actions together tasks which are complemented by other secondary activities such as long distance reconnaissance, re-supplying, medical evacuations "MEDEVAC" (MEDical EVACuation), the transport of personnel etc.

The Introduction of the helicopter

After the second world war, the process of introducing helicopters began with the

MULTINATIONAL
The NH90 helicopter is the fruit of a multinational project to produce a family of advanced naval helicopters which could satisfy the specific needs of the marines of France, Italy, Germany and The Netherlands.

idea of them being appropriate for various phases of combat, including naval missions. The need for these machines to supply the marines of various European nations led to their co-manufactuer in the 1950's, under a United States design licence. This was an advantageous experience with the modernizing of models and the creation of new machines adapted from the needs highlighted in the development stages.

ADAPTED
The "Lynx" helicopter has been adapted to carry out naval activities for which it has received very advanced specialized equipment which has favored it in exports markets.

Evolution

The British company Westland obtained a licence in 1959 for the manufactuer of the United States Sikorsky S-61/SH-3 SEA KING. The company began to manufactuer and supply them to the Royal Navy as a heavy helicopter capable of executing anti-submarine tasks. Developing it through the years into specialized versions for use in anti-submarine warfare (ASW) and in an airbourne early warning role using the Thorn-EMI Searchwater radar and as transporter for the Royal Marine comandos. Its manufactuer has exceeded 325 units, it is also employed by the marines of Germany, India, Pakistan, Australia, and Egypt.

Later, the Italian company Agusta Spa obtained the license to manufacture the 212 model as an anti-submarine variant. This was as a result of its experience in constructing, under license, the Sikorsky SH-3 and various models of the United States Bell Company. Known as the AB-212, this twin-turbine model incorporates modified avionics and different equipment to allow it to act as a submarine hunter or anti-ship helicopter, of which more than a hundred units have been manufactured for the marines of Italy, Greece, Turkey, Spain, and Venezuela. Spain has incorporated a dozen of these including four which have been modified to carry out electronic support missions using the Colibri system.

JOINT MANUFACTURED

The LYNX was born as the result of the relationship between British and French companies, and is usually deployed on French frigates and destroyers, equipped with a landing pad and maintenance hanger.

WIDESPREAD

The Agusta Bell 212's have been exported to numerous countries including Spain which has incorporated a dozen of them in its Third Air Force Navy Squadron for transport, electronic warfare, and naval surface and search missions.

Own capability

Previous developments, along with other important aspects, brought about the birth of their own designs which benefited from previous manufacturing experience. Together with the process of optimization and adaptation of the models in service. The first to be conceived was the Lynx which was born as the fruit of a collaboration between France and Great Britain- 30% the former and 70% the latter. Entering production in 1976 after nine years in development, the Lynx has since been adopted in its specialized naval form by Germany, Great Britain, France, Portugal, Brazil, Argentina, Egypt, The Netherlands, Denmark, South Korea, and Nigeria, with more than a hundred machines in service. The Lynx and the Super Lynx specialize in A.S.W. air-to-surface actions. The latter of the two being an updated model equipted with the most modern and advanced technology.

In 1979 an agreement was signed between Westland and Agusta for the development of a new helicopter to replace the Sea King and the lynx. After creating a joint company known as European Helicopter Industries. They began the work of constructing the EH-101 Merlin which flew for the first time in October 1987. The first units entered service in 1998 in British and

Their details

Each one conceived at a different time and with the objective of meeting different operating requirements, their characteristics differ in size, mission possibilities, equipment and tactical use.

The "Sea King"

Manufactured in Italy and Great Britain, it has the capacity to hunt and destroy its target which allows it to operate independently from surface ships. It was conceived around a large core which included a cockpit at the front for the pilot and co-pilot, a lower part which allows it to make emergency landings on water, two lateral elements which are home to the main landing carriage, which is retracted during flight, and a large free area where up to 22 people can travel and also where different kinds of equipment can be put.

The British machines are equiped with a GEC-Marconi AQS-902G-DS processor associated with the 2069 sonar, a sonar buoy launcher, AN/ASQ-50 magnetic anomalies detector, a display screen for the systems operator, surface search operations are carried out by an integrated navigation system whose main component is the Thomson Thron AR15955 Radar.

Powered by two Rolls Royce Gnome H-1.400-IT Jet Engines which together give

Italian ships. There are as of yet sales to other countries such as Canada or Spain, which could substitute these for existing helicopters.

Other specific versions have been produced for naval use like the SUPER FRELON, the Eurocopter AS 365N3 Dauphin 2, or the NHI Industries NH90 destined to equip future frigates for France, Italy, Germany and The Netherlands.

TECHNICAL CHARACTERISTICS

	AB-212 ASW	"Sea King"	"Lynx"	"Merlin"
COST IN MILLIONS OF DOLLARS:	12	14	22	35
DIMENSIONS:				
Fuselage length	17.4 m	22.15 m	15.165 m	22.81 m
Height	4.53 m	5.13 m	3.48 m	6.62 m
Main rotor diameter	14.63 m	18.90 m	12.80 m	18.59 m
Main rotor turning area	173.9 m²	280.48 m²	128.71 m²	271.51 m²
WEIGHTS:				
Empty	3,240 kg	5,447 kg	2,740 kg	7,121 kg
Maximum	5,070 kg	9,752 kg	4,876 kg	14,600 kg
Maximum arms load	800 kg	1,300 kg	700 kg	960 kg
Fuel	814 l	3,714 l	957 l	3,222 l
PROPULSION PLANT POWER	1,875 CV	3,320 CV	2,240 CV	6.936 CV
PERFORMANCE:				
Ceiling service height	4,023 m	1.705 m	3,230 m	5,000 m
Maximum speed	196 km/h	272 km/h	232 km/h	278 km/h
Range	667 km	1,482 km	500 km	900 km

the 3,320 horsepower necessary to give agility to the main five-bladed rotor, it can be fitted with a BL10300 side winch with the capacity to hoist loads of up to 272 kg, or with a varied range of launchable weaponry. In the case of the British units this include up to four STING RAY search torpedoes, four MKII depth charges, two Sea Eagle anti-ship missiles, and a medium support machine gun installed in a lateral mount, which is operated from inside the helicopter.

The AB-212 ASW

Manufactured in Italy by Agusta, this derivative of AB-205 for naval use incorporates various changes to the core which allow it to be adapted to different operating requirements. Important for working over the surface of the sea is the Pratt & Whitney Canada PT6T-6 Turbo Twin Pac, a tin twin jet engine which produces 1,875 horsepower with reduced consumption, which allows the machine to stay in the air for up to 3 hours.

The specialized equipment of these Italian units consists of a MM/APS-705 surface search radar located in a cylindrical housing above the cockpit, an AQS-18 sonar unit of variable depth and low fre-

ESPECIALIZATION

The latest versions of the British Super Lynx are equipped with the most advanced electronic systems including a thermal imaging camera, surface search radar and a complex front assembly which differentiates it from other members of this family.

JOINT MANUFACTURED

The LYNX was born as the result of the relationship between British and French companies, and is usually deployed on French frigates and destroyers, equipped with a landing pad and maintenance hanger.

quency and an operator screen. With a reinforced core for improved resistance to marine corrosion, skids supplied with floats for emergency landings on the sea surface, and strengthening for berthing under cover. It has surpassed other more modern designs and has been consigned to various fleets, such as Spain's. It carries out other secondary missions such as transportation, liaison work, surface searches, and tasks involving the use of rocket launchers, Whitehead A-244/S anti-submarine torpedoes, mines or anti-ship missiles such as the Marte MK2.

The "Lynx"

Compact and with great agility in the air, this small British-French helicopter is based on a one-piece fuselage constructed of light alloy, with doors, tailplane and access panels made from glass fibre. Its characteristics are: a three-wheeled non-retractable undercarriage, a cockpit equipped with the most advanced equipment which gives it a high capability, a compact propulsion plant consisting of two Rolls Royce Gem2 Jet Engines with a combined power of 2,240 horsepower, and in the case of one of them failing, a system to increase the power in one to the maximum possible. Capable of carrying 10 men or for storing equipment particular to the navy.

The British helicopters have (GPS) global positioning systems; GEC-Marconi A15979 Seaspray MK1 Search Radar with the capability to locate small targets in conditions of low visibility and rough sea conditions, Bendix sonar equipment; Sea Owl thermal imaging cameras; radar emission detectors etc. The French units are armed with Mk46 and Murene light torpedoes, depth charges and AS-15TT anti-ship missiles.

The "Merlin"

Belonging to the third generation, the EH101 is a large helicopter for those fleets which require a wide range of possibilities for search activities associated with their defence. Created mainly with an aluminium and lithium fuselage, composite materials, and sandwich type panels. Its missions are mainly to search for and neutralize surface naval units and submarines, for which the

COMPLEX

The complexity of the equipment installed in the anti-submarine helicopters means that highly qualified personnel are required to do the maintenance and installation work-done in the stern hangers of frigates and destroyers.

HEAVY DUTY

The French "Super Frelon" has carried out multiple naval missions, and amongst which is the noteworthy anti-ship hunter version equipped with powerful "EXOCET" missiles located in lateral mounts.

fuselage has been developed so that it is big enough to be fitted out with the specialized equipment and weaponry necessary. At the same time it is small enough to be able to fit into the small hangers of frigates.

It is capable of operating from its carrier ship in winds of up to 54 mph and in sea conditions of gale force 6. Incorporating an automatic hook up system to the flight deck. Its main rotor can adopt a negative angle of incidence to facilitate landing in extreme conditions, in which the five-bladed main rotor is very useful, driven by Rolls Royce/Turbomeca RTM322 Jet Engines, producing a combined power of 6,936 horsepower.

Installed in Martin Baker Armored Seats, two crew members work in a cockpit which includes six display screens in Litton Color, while the operator of the acoustic surveillance system and observer travel in the cabin behind, which is also equipped with a display console with four monitors. Thanks to this it is simple to control the GEC-Marconi Blue Kestrel Search Radar, the AQS-903 acoustic processor associated with the Thomson-Sintra Sonar System which extends to a depth of 600 metres. The Racal Electronic Components Support System which identify and locate hostile radar, the JTIDS coded link; the attack equipment which, with a maximum load of 960 kg, includes four light search torpedoes, two anti-ship missiles, depth charges, and machine guns for self-defence decoy to confuse the heads of guided missiles.

S pecifically designed to be used as an armed helicopter during the Vietnam War, the AH-1 COBRA and its updated version SUPER COBRA have demonstrated great versatility and high potential. This has motivated the creation of improved versions that it is thought will remain in service until the end of the second decade of the 21st century.

Very agile and maneuverable, with the incorporation of a main rotor with only two blades, it has been widely used in the conflicts in Vietnam, Lebanon, Grenada, Panama and the Gulf War. In missions in these conflicts its robustness and potential as a combat unit were well demonstrated, despite

OPTIMIZED

Although the AH-1 has been in service for many years, the modern Cobra family includes optimized versions that have been fully updated to conform with the elements of modern combat.

some having been shot down for being more vulnerable than other more modern designs which incorporate better physical and electronic self-defence elements.

The need is created

It was first produced in June 1962, the fruit of a private corporate development known as the D-255 Iroquois Warrior. It also incorporated some elements developed for the Bell UH-1 Utility Helicopter. The requirements of the United States Army for the AAFSS program (Advanced Aerial Fire Support System) necessitated it development into a concept known a Model 209, the prototype of which flew on the 7th of September 1965.

The selection of model 209

This was selected on the 7th of Apri 1966, when two test models were contracted that were followed six days later by 11 production machines; the first AH-1G were sent to Vietnam in August 1967. The mission needs in this conflict, in which several hundred of these machines were

CANNON

There is a General Electric GTK4A/A assembly installed underneath the front part of the fuselage which includes a three-barrelled M197 gun mounting with three 20mm cannons, capable of firing limited bursts of 16 rounds at a maximum rate of 675 rounds per minute.

shot down or suffered accidents, prompted the Marine Corps to adopt the machine in a better designed twin-engine model, designated the AH-1J Super Cobra, that was ready in 1969.

The introduction of new capabilities, such as the launching of TOW anti-tank guided missiles, the incorporation of more powerful engines, its provision with improved avionics and other small details have led to the construction of two thousand machines. These have also been acquired by other nations, among which are Israel, Turkey, Greece, Iran, Jordan, Pakistan, Spain, Bahrain, South Korea, Thailand, and under licence in Rumania and Japan.

The Cobra family

The large range of Cobra helicopters produced includes a long evolution of models adapted to the necessities of the moment. The first prototype of Model 209 incorporates a retractable undercarriage on the underside of the fuselage that was substituted by a fixed unit with two skids on the production models. The first model to enter service was the AH-1G, of which 8 models were destined for the 7th Armed

Air Squadron of the Spanish Navy, followed by advanced models such as CONFICS, ALLD, ATAFCS or SMASH, depending on the tracking systems installed.

The former was followed by the AH-1Q with the capacity for launching TOW missiles; the AH-1S with more powerful engines that increase its agility and maneuverability; the AH-1P with flat walls on the sides of the cockpit and warning radar; the AH-1E with 20mm cannon M197 mountings and the AH-1F, optimized for anti-tank combat.

The United States Marines have also acquired twin-engine models designated "Super Cobra", which include the models AH-1J and AH-1T, modified in order to gain more maneuverability and improve its

chances in air to air combat. There is also the AH-1W which corresponds to the most powerful version currently in service and which has now been programmed to be modernised up to the standard AH-1W/4BW models. These incorporate General Electric T700-GE-4101 jet engines with infrared suppressers, new avionics, a high-resolution FLIR tracker, improved stub wings, and more. These transformed models are set to be handed over to the forces between 2004 and 2013.

The Super Cobra

Designed by the USMC and bought by Turkey and Romania, the AH-1W "Super Cobra" is the most powerful model of the Cobra family designed to date. It stands out as much for the survival possibilities the two engines give it, as for the high agility derived from the employment of a two-bladed propeller for the main rotor.

Characteristic

With a similar fuselage in all the models of the range, this helicopter presents a notably narrow profile that makes its

upper part is the main rotor which move a large double-bladed propeller and whic is driven by two General Electric T700-GE 401 jet engines, capable of generating combined power of 3,250HP. They are pro-vided with a large cover, which facilitate maintenance tasks. The rear section i elongated and includes the small tail roto with a 2.94m diameter propeller which ha been manufactured in combined aluminiun and steel.

To optimize the work of the crew, an their capability to face the missions entruste to them, there is a Kaiser Data Display fo

detection from the front difficult and reduces its radar and infrared signature. The cockpit is the tandem type with the pilot seated in the rear section, raised-up to obtain better visibility of what is happening outside, and the co-pilot systems operator located in the front part, from where he uses the viewfinder that allows aiming and firing. Both incorporate displays compatible with use of night goggles, and have lateral and underside protection against light-arm impact, and enjoy air conditioning.

The central area of the front section incorporates two skids on the lower part that allow landing on any surface. On the

the pilot, AN/APN-194 altimeter radar AN/APN-44(V) radar warner, AN/ALE-39 interference flare launchers, Teledyne AN/APN-21 navigation radar based on Doppler impulses with Collins display screens, and coded communication links.

Potential

The inclusion of an M65 viewfinder, modified with an NTSF-65 thermal imaging camera from the Israeli Company Rafael, give it the capacity for daytime and night time operations and for the use of advanced weaponry. Notable amongst these are the GTK4A/A General Electric nose turret

which includes a three-barrelled M197 gun-mounting with three 20mm cannons of 1.52 meter length that are capable of firing limited bursts of 16 rounds at a rate of 675 rounds per minute. Their feed drum is capable of housing 750 rounds. The firing angle is 220° side-to-side, 50° downwards and 18° upwards.

On the side stub wings 70mm rocket launchers can be positioned that range from the LAU-69As with 7 tubes, to the LAU-61As with 19 tubes CBU-55B housings of air-combustible explosives; SUU-44/A flare-launchers, M118 grenade-launchers, GPU-2A or SUU-11A/A containers for 7.62x51mm multi-barrelled Minigun machine-guns, TOW or AGM-114 "Hellfire" anti-tank missiles, Hughes AGM-65D "Maverik" surface-attack missiles, or air to air AIM-9L "Sidewinder" missiles. A very wide range of weapons that give it enough power to take part in escort missions, armed reconnaissance, target identification or attack against mechanized or armored vehicles.

In combat

With respect to the use of arms it has been demonstrated that the Israeli AH-1Ss have used the "Hellfire" missiles to attack the buildings of the headquarters of the fundamentalist Hezbollah group and the Palestinian positions in the south of Lebanon. Meanwhile the United States machines had the highest operational readiness of all the helicopters that took part in the operations to free Kuwait. One operation that particularly stands out was the neutralization of the first armored units of a column of 1000 vehicles on a bridge, using TOW missiles. Once immobilized they were easy prey for the rest of the weapons used by the international coalition force.

MAINTENANCE

Conceived for combat, the Cobra requires specialist maintenance to tune up the multiple systems that give it its capacity for attack missions.

MISSIONS

Identified by the initials SFOR on the fuselage, these Cobras have carried out armed reconnaissance missions over the former-Yugoslav skies during the peacekeeping phase assigned to NATO.

TECHNICAL CHARACTERISTICS AH-1W

COST:	10.7 million dollars		Maximum load arms and fuel:	2,019 kg
DIMENSIONS:			Internal fuel	1,128 l
Length	13.87 m		**PROPULSION:**	TWO GENERAL ELECTRIC T700-GE-401 JET ENGINES,
Height	4.11 m			EACH UNIT WITH A POWER OF 1,625 HP
Rotor diameter	14.63 m		**PERFORMANCE:**	
Rotor turning area	168.11 m²		Service ceiling height	4,495 m
WEIGHT:			Maximum speed	352 km/h
Empty	4,671 kg		Cruising speed	278 km/h
Maximum	6,690 kg		Range	587 km

CO-PILOT

The gunner/bombardier co-pilot has for his use a view-finder connected to the front tracking turret. He is in charge of firing the arms systems with the side lever, which includes a red-colored trigger.

SENSORS

A stabilized housing contains the navigation and guidance sensors of the AH-1, consisting of a thermal imaging camera, television camera, laser rangefinder equipment and so on. These components allow it to operate equally well by day as by night, and in adverse weather conditions.

CANNONS

The M197 mounting incorporates three 20mm cannons that can fire bursts of up to 3,000 rounds per minute and enjoys a wide angle of fire in order to confront land-based or aerial targets.

ROTOR

Developed from that of the Bell UH-1 helicopter, the principal rotor head is articulated, turns clockwise and has shown itself suitable to give the Cobra a high level of flight agility.

TWIN JET ENGINES

The Super Cobra version of the Marines includes 2 General Electric T700-GE-401 jet engines that produce a total of 3,250HP, with which the helicopter can perform its combat missions without restriction equally over open sea or land.

TAIL

The rotor is incorporated in the upper part of the tail structure, which has the function of maintaining directional stability. There are right-angle section stabilizers on the sides and a skid underneath that protect the tail beam against possible blows during take-off or landing.

STUB WINGS

Each of the two stub wings consists of two rigid fixings for TOW anti-tank missiles, rocket launchers, incendiary liquid tanks, a variety of missiles, transport containers and other systems. Recently, a magazine has been installed on top of the wings for launching interference flares.

SKIDS

Underneath the main structure are the two skids, one on each side, that allow the machine to operate from any surface and have shown great durability throughout its service. To facilitate movement on the ground, a pair of small wheels can be fixed to the front parts.

Designed as a helicopter with the ability to neutralize multiple surface targets, the "Apache" had its baptism of fire in the Gulf War where its features were demonstrated in the destruction of a good part of the Iraqi mechanized columns and armor, as it also did with the neutralization of other targets such as command posts, anti-aircraft systems, communication systems and a long list of others which gave us proof of its capability as a combat vehicle.

Facing the threat

The importance of the large and growing concentration of armored vehicles in the now extinct Warsaw Pact, was the reason for the United States developing an airfleet to break its hypothetical advance.

The Conception

After Lockheed´s "Cheyenne" AH56A program was cancelled and the attack missions during the Vietnam conflict were left to the Bell AH-1 "Cobra", which demons-

> **THE LONGBOW APACHE**
>
> Chosen by the armies of the United States, United Kingdom and the Netherlands, the AH-64D is the most modern version of the "Apache" multi-mission helicopter and incorporates various improvements in a number of its sensors, its weapon capacity and cockpit design, increasing its ability to act when facing all kinds of targets.

> **THE CAVALRY**
>
> Various squadrons of United States cavalry use "Apache" attack helicopters for operations from carrying out reconnaissance missions to others of mass destruction.

trated weaknesses when carrying out some missions, the requirements for the AAH (Army Advanced Attack Helicopter program were formulated. After investigating various proposals a contract was signed on the 22nd of June 1973 with Hughes Helicopters and Bell Helicopter Textron which respectively proposed their YAH-64 and YAH-63 models, of which the first would be chosen, going into flight on the 30th of September 1975. Following a long period of development time, and after modifications to constituent parts, the first model of the series left the factory in September 1983 and was delivered to the United States Army the next year; on the 26th of January 1984.

Production

Contracted in the 1982 budget, the first 11 operational models were added to the 6 prototypes (one for ground trials, and five flight-worthy models), the deliveries allowing the needs to be met of the crew training center in Fort Rucker (Alabama) and also those of specific training centers in areas such as maintenance, logistics, avionics etc. The third squadron of the 6th cavalry regiment became the first unit to reach operational readiness in July 1986.

After orders for a total of 827 "Apaches" and a production rate that has allowed the manufacture of more than a hundred machines a year, the last (the 'A' model) was delivered on the 30th of April 1996 to one of the 35 air cavalry squadrons equipped with this model. Others also with this model include the National Guard, with seven, and the Reserve with two. The decision to increase the

number of helicopters in each battalion from 18 to 24 has meant that from 1997 there are 26 units operating with the AH-64A.

The Improvements

Despite the improvements made to the original design, such as the ability to employ the M230 cannon ("Chain Gun") in air to air combat. The possibility of employing "Stinger" or "Mistral" infra-red missiles for self defence against other armed equipment or to shoot down slow flying planes and helicopters; the introduction of a global positioning system (GPS); anti-interference radios ("Singars") or the adoption of an integrated electronic warfare system, it was decided in August 1990 to begin updating the helicopter, resulting in the AH-64D Longbow Apache.

The features of the new version have been validated by evaluating the working of various prototypes, of which the first flew on the 15th of April 1992. An agreement was made in 1996 to make 232 machines up to the year 2000 and in March 1997 the delivery began of the first AH-64Ds manufactured by McDonnell Douglas, the company which now manages production. A period of ten years is forecast to transform the 758 machines that make up the existing fleet.

Capacity

After requests for more than two hundred new or army surplus models from Israel, Saudi Arabia, Egypt, Greece, The United Arab Emirates, The Netherlands or The United Kingdom, which manufactures them under licence, the forecasted number to be built is 1,040 which can be added to requests from a variety of other countries showing an interest in the United States attack machine. These countries include Kuwait, Malaysia, Singapore, Sweden, South Korea or Spain, which were hoping to acquire thirty attack helicopters for its Helicopter Attack Battalion (BHELA).

An advanced design

With a robust configuration that improves its performance and capacity the "Apache" has been designed on advanced lines that increase the likelihood of the survival of the pilot and co-pilot/weapon operator. The range of possible weapons has increased, it has a high resistance to impacts from light weapons and anti-aircraft fire and it acquits itself well in accidents or ground crashes at speeds up to 12.8 metres per second. This is thanks to its landing carriage or undercarriage which has a high capacity for absorbing impacts.

Both crew members travel in a heavily armored double cockpit, which resists armor piercing missiles of dimensions up to 12.7x99mm and they have at their disposition a variety of systems to carry out their activities, such as flight helmets with integrated data displays and guidance systems, multi-functional cockpit display screens and a wide range of equipment that includes the

AH-64D LONGBOW

Doppler Plessey AN/ASN - 157 navigator, a digital auto-stabilizer, and a navigation system that combines a radar altimeter, laser, aerial data, INS, Doppler, and GPS.

The 1,890 horsepower of each of the two GET700-GE701C jet engines, mounted on each side of the fuselage give it great agility and the ability for the helicopter to acquit itself well in multiple operations. The jet engines are covered with armored plating ,which act as maintenance platforms. The radius of operation is limited by the fuel, a total of 1,421 litres that can be carried in four additional external tanks, from the Brunswick Corporation, with a capacity of 871 litres each. With a high level of protection a helicopter can fly for 30 minutes after being hit by 12.7mm projectiles in any part of the fuselage, although in some zones it can resist 23mm. Its transmission can continue working for an hour even though there is no lubricating oil.

Combat
The most modern version of this Uni-

> **SPECIALIZATION**
> Its power, capability, and possibilities of attack have made the AH-64 the reference by which later attack helicopter designs are measured.

> **CANON**
> Installed below the fuselage, guided automatically towards its objective by sensors and fed by a considerable ammunition supply, the 30mm M230 canon, the "Chain Gun", can fire at a rate of 625 rounds a minute and its effectiveness is striking against all types of armored vehicles and targets.

ted States helicopter is capable of working equally well at night as it is during the day, or in adverse weather conditions, for which it is equipped with laser, infra-red and other high technology systems that allow it to detect, classify and prioritize targets. It then tracks and attacks them. The first activities involve the use of the Westinghouse Radar which is capable of presenting up to 256 targets on the tactical positioning screen. This is mounted on the mast above the main rotor. This is an acquisition system from Lockheed Martin Orlando Aerospace with the joint TADS/PNVS which has an infra-red sensor

presenting a thermal image of the point being observed and an integrated counter-measure system- SN/ALQ-211SIRFC (Suite of Integrated RF Countermeasures) which complements the warning alarm sys-

tem for the detection of radar and laser emissions, the infra-red interference system AN/ALQ-144 and rocket launchers M-130 acting as decoy flares. British helicopters substituted the SIRFC for a GEC-Marconi HIDAS system (Integrated Defence Aids System).

Attack missions are entrusted to the McDonnel Douglas automatic canon, the machine gun, which is capable of firing 30mm bullets at a rate of 625 rounds per minute, fed by a magazine of 1,200 rounds. There are two stub wings with four attachments which can hold 16 Hellfire anti-tank RF missiles, 2.75 inch flare rockets, air to air missiles with infra-red guidance systems, Sidearm anti-radar missiles etc. All of these allow it to act as an armed escort, hunter or attack helicopter.

PROPULSION

Installed on the sides of the fuselage and with quick release panels to facilitate maintenance work, the two GE T700- GE-701C jet engines together deliver 3,780 horsepower and incorporated in the outlet there is a nozzle which reduces its own infra-red signature making it difficult to locate.

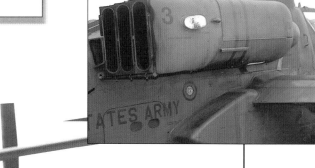

ROTOR

The asymmetric four-bladed tail rotor is located on the left side of the machine in a way which optimizes and stabilizes the agile movements of this attack helicopter.

SHOCK ABSORPTION

A small wheel is incorporated below the tail with a shock absorption system which prevents the back section of the fuselage from touching the ground during hard landings when being employed in combat duties.

WINGS

On the side of the fuselage there are two stub wings with two attachments for the location of weapons like Hellfire anti-tank missiles, rocket launchers or auxiliary fuel tanks.

RADAR

The Longbow variant incorporates a Westinghouse radar in a semi-spherical housing located above the main rotor, with the function of locating targets and optimizing the guidance of weapons in the attack.

COCKPIT

The AH-64D has improved the work possibilities of the pilot and co-pilot with the implementation of cockpit control systems which now give multi-function display screens and have increased the time available for doing other activities.

SENSORS

A stabilized gyro is located in the front nose which guides the thermal sensors and laser during flight, including in adverse conditions, enabling the use of the onboard weapons.

AH-64D TECHNICAL CHARACTERISTICS

COST:	18 million dollars	Fuel, external	3,484 l
DIMENSIONS:		**PROPULSION:**	
Length	15.47 m	Two GE T700-GE-701C jet engines with a total	
Height	4.95 m	of 3,780 horsepower	
Width	5.227 m	**PERFORMANCE:**	
Main rotor turning area	168.11 m²	Service ceiling height	6,400 m
Tail rotor turning area	6.13 m²	Stationary ceiling height	4,115 m
WEIGHT:		Maximum speed	261 km/h
Empty	5,352 kg	Range	407 km
Maximum	10,107 kg	Extended range	1,899 km
Maximum external load	2,712 kg	Design factor loading	+3.5/-0.5 g's
Fuel, internal	1,421 l		

MULTIPLE

Its robustness, weaponry capacity and armor-plating make the Mi-28 capable of multiple missions in every kind of battlefield and in the most adverse meteorological conditions.

Conceived by the Soviet industry as a technological reply to the United States Apache, the Mi-24 was created to complement the powerful Hind. It comes from a long tradition of manufacturing by the company Mil which has built an excess of 31,000 helicopters. The continual lack of funds has impeded its construction of large runs and its construction has depended on different sources. Seventy units are in active service with the Russian Federation Air Force.

Creating a need

Despite the fact that the Soviets had exported a lot of the Hind Helicopters to many of its orbital countries (political and economical) the continual advances introduced by western countries, with designs like the Italian Mangusta, the United States Cobra and Apache, made Russian strategists decide to create a new model which would include everything a modern attack helicopter needed. This was to be supplied to their own armed forces in a way which would also act as a possible export channel.

Development

Taking as a reference the project carried out by Alexei Ivanov, the initial design work began in 1980 in the Mil OKB Factory. The first of the four prototypes had its first flight on the 10th of November 1982. In the 1989 Bourget Airshow the third of these was presented to the western public having just completed 90% of the development process after 800 hours of flight. Produced by Rostvertol in Rostov, its introduction was very slow and export orders were not arriving,

Missions Types

Although the philosophy for the deployment of helicopters in Russia has gone through changes after some experimentation, it is being used at division level in ground units where it is favored for its quick utilization and deployment through an effective chain of command.

Missions trusted to the helicopter would involve tasks such as close support offensives, attacks against armored formations, eliminating enemy anti-aircraft defences, anti-helicopter defences, rearguard operations and general support. It operates together with fixed-wing planes such as the Su-25 "Frogfoot" and its variants.

These tasks are carried out in three zones. The first is a band of 15 kilometres depth inside enemy lines and in this area it carries out anti-tank missions and neutralization of missile and electronic defence systems. The second extends up to 30 km and in this area supports the operations of its own side's airplanes; meanwhile the third covers up to 150 km and includes deep penetrations to accompany infiltration helicopters and rescue missions for pilots who have been shot down.

A design with special features

With a very similar design in much of its conception to its western equivalents, the United States Defence Department declared that in principal the HAVOC was a copy of the APACHE. However this machine presents a variety of interesting peculiarities as a result of the different conceptions of the industry which produced it and which was looking to combine robustness with its other features.

---

although it was being evaluated in Sweden. Despite this, from January 1994 it was being worked on to meet the requirements of the Mi-28N concept; which is capable of operating at night and in bad weather. This was the reason for the radar mounted above the rotor which could be seen in the last part of the Bourget Paris Airshow.

Different versions

The initial forecast was to construct up to six different versions of which the anti-tank series would occupy the greater part of the production capacity. It would also be used to replace the oldest HIND models. The basis of the development would allow the evolution of a model with greater capacity to operate in any weather, satisfying the needs of a variety of nations interested in robust, capable Russian designs.

Another version being worked on is for the navy, with protection systems to work in this kind of environment. Designed to operate from amphibious vessels giving marines cover when they are disembarking and, at the same time, being able to confront threats such as patrol boats.

Finally, let's describe the capabilities of specific versions optimized for air-to-air combat with other similar helicopters and with the ability to face aeroplanes in slow and low flights and others designed for anti-missile purposes.

HAVOC

With a low rate of production due to the continual reduction in the Russian defence budget, the Mi-28 is a machine conceived to replace other older models which are out of step with newer western models.

CANON

Integrated at the front and below the fuselage, the 30mm cannon is fed by two ammunition magazines and is capable of controlled bursts of fire of up to 900 rounds per minute with an effective reach of more than two kilometres against surface or air targets.

The structure

It is robust and heavily-armored, the cockpit includes mechanical flight controls moved with hydraulic assistance and the latest development sees the installation of liquid crystal display screens (LCDs). From the cockpit to the main rotor which has five blades receiving rotary power from two Kimov TV3-117VM Jet Engines which produce 2,070 horsepower each. They are located in lateral gondolas with infra-red suppressers in the outlets that the signature is 2.5 times less than its predecessor the Mi-24. Between both Gondolas is a small Ivcheknko AI-9V Jet Engine which acts as an auxiliary power unit to start up the propulsion units if they need external assistance.

The fuel tanks which supply the propulsion units have a capacity of 1,665 litres and are filled with a polyurethane foam designed to avoid accidental explosions. In its interior it is covered with a layer of latex which seals holes produced by projectiles or shrapnel.

POWER

A radar fitted above the main rotor, a complex radio guidance system located in the front section, launchers at the extremes of the wings where the weaponry is located, a 30mm cannon at the front, and a long list of etceteras for this powerful Russian attack machine.

At the height of the entrance nozzles for the turbines we find the components which make up the main landing carriage which are notable for their high-capacity for absorbing the shocks of impacts up to 17 meters per second.

A design for survival

Designed for the highest level of survival, the cockpit and main features of this

NOZZLES
The two KLIMOV TV3-117VM jet engine plants stand out for their power of 2,070 horsepower each and for their outlets pointing downwards to reduce the thermal signature. There is a small turbine between the larger ones which allows autonomous operation without external assistance.

COCKPIT
With good visibility and strong armor plating the cockpit is notable for its advanced equipment, but which is somewhat inferior in its performance and modernity to those installed in the western equivalents.

CAPABILITY
After the latest systems incorporated the Mi-28N has been converted into a powerful and capable attack machine which is only waiting for the confidence of other countries for it to begin export sales (The Swedes have been evaluating the helicopter).

machine are heavily armour plated against every type of light arms attack with components of titanium and composite materials specifically designed for such use. The cockpit seats have a pyrotechnic system which tenses the harnesses in the event of impacts and are provided with parachutes in case of emergency. Positioned in these seats the pilots enjoy the protection of a cockpit which resists 20mm impacts and flat glass windows which are designed to cope with 12.7mm projectiles. The whole front part resists light arms fire from distances of 30 meters.

The fixed-landing carriage is very robust and the cockpit floor employs panels which follow the bee-hive model giving it the capability of surviving a vertical fall of up to 17 meters per second without any of the crew coming to any harm.

Capability

Usually, flying at altitudes less than 20 meters avoids neutralization, and this machine incorporates in its front section a complex monitoring and flight support system which includes a small radar transmitter for guiding missiles, an image intensifying television camera, a laser transmitter and soon it will incorporate a FLIR infrared tracking system.

These advanced and extremely effective components, which are connected to

its own navigation system, give it the possibility of employing its own weaponry in a way which is able to give it a much greater guarantee of success. The 30 mm 2A42 cannon is mounted in the turret which can be found just below the nose at the front of the helicopter, and is fed by two magazines with a capacity of 250 rounds. In addition there are two fixtures for more weapons located below the stub wings which can be used to hold infra-red rocket launchers. These mounts are capable of carrying up to 480 kilograms of arms.

A standard configuration for this helicopter could be considered as one which includes 16 AT-6 Spiral Anti-Tank missiles, or instead could involve using the new Ataka System. This is a launcher where the missiles have a maximum range of 8 kilometers, or an effective range of 5 km, and when non-guided up to 80 km. Yet another alternative is to use 122 mm rocket-launchers for what is a different kind of mission, in air-to-air combat whith other enemy aircraft.

HAVOC MI-28 TECHNICAL CHARACTERISTICS

COST IN MILLIONS OF DOLLARS:	Unknown
DIMENSIONS:	
Length, excluding rotors	17.01 m
Height	4.70 m
Wingspan	4.88 m
Diameter of main rotor	17.20 m
Main rotor turning area	232.35 m²
WEIGHT:	
Empty	8,095 kg
Maximum	11,660 kg

Internal fuel	1,665 kg
PROPULSION:	
Two Klimov TV3-117VM jet engines capable of producing together 4,140 horsepower	
PERFORMANCE:	
Service ceiling height	5,800 m
Maximum speed	300 km/h
Tactical operational range	200 km
Reach	1,100 km
Design factor	+3/-0,5 g's

COCKPIT

The glass covering the cockpit of the pilot and co-pilot/gunner is designed to avoid revealing reflections and is thick enough to absorb the impact of weapons up to 12.7mm caliber.

SENSORS

Radar for guiding missiles, laser rangefinder and an infra-red tracking module are some of the sensors integrated in the front part of the Mi-28 which give it a high-altitude flight position in the launching of its diverse weapons package.

LANDING CARRIAGE

The major elements of the landing carriage incorporate a shock absorption system which cushions the fuselage in ground impacts up to a velocity of 17 meters per second and without which the crew would be injured.

ROTOR

The radar ,installed in a housing above the main 5-bladed rotor, has improved the possibility of employing the aircraft in whatever kind of meteorological conditions.

TAIL

The tail part has a 4-bladed rotor which improves directional stability and also has a small wheel which, in this area, softens impacts against the ground on landing.

STUB WINGS

The stub wings support the weapon load with, at their extremities, housings which launch interference flares and underneath there are two robust supports, one incorporating 8 ATAKA anti-tank missiles and the other twenty 80mm rockets.

FUSELAGE

Incorporated in the upper part of the fuselage are enemy tranmission detectors and communication aerials and it is highlighted that additional equipment or personnel can be carried in this area.

United States naval strategists wanted to achieve dominance of the seas, especially when taking into account the important growth of the Soviet Submarine Fleet. This led them, in the 1950's, to begin a strengthening program which principally

The introduction of different models

With the objective of carrying out the needs of specialized machines, various models entered service from the beginning of the 1950's, some of the better known ones being the Piasechi HUP-2S Retriever, the HRS-2 and HSS-1 Seabat. After experimenting with these designs the United States Navy, in December 1957, contracted the firm Sikorsky for the manufactuer of ten units of an advanced design, designated YHSS-2, the first of which flew on the 11th of March 1959, resulting in the SH-3A Sea King-equally capable of hunting and destroying in flight.

resulted in the establishment of a powerful group of aircraft carriers to carry squadrons to wherever they were needed.

To protect them and to provide coverage for transport ships, it was planned to put into service escort ships such as frigates and destroyers, supplied with helicopters which were specialized in the work of detecting and destroying enemy submarines.

Acquisition

SEA KINGS were equipped with a Bendix AQS Sonar System and Ryan APN-130 Doppler Radar managed by two operators located in the cabin. In parallel to the first 254 units contracted and entering service, there were also the first Kaman HU-2A Seasprite light helicopters, which were assigned to the role of navy search and rescue.

After removing the ASW aircraft carriers from the ranks of the US Navy it was decided to optimize the SEA KING. In 1967 the modernized SH-3D version entered service with more powerful engines and more modern systems, such as the AQS sonar and APN-182 Doppler Radar. The introduction of the LAMPS (Light Airborne Multi Purpose System) came about by the contracting of twenty modified Seasprite helicopters at the beginning of the seventies in the anti-submarine SH-2D standard. Incorporating a Canadian Marconi LN-66HP Surface Radar located below the cockpit, a Texas Instruments ASQ-81 Magnetic Anomaly Detector (MAD), a 15 tube launcher of sonar buoys, active and passive. A mounting for the carrying of two light torpedoes, all of which are controlled by a system specialist.

Improvements introduced

Later and up to 1982, another hundred of the standard SH-2F units were modified.

"SEASPRITE"

Two United States Navy Fleet Reserve squadrons continue using the SH-2F in submarine detection missions and general naval support, missions which will continue up to the middle of the next decade.

ANTI-BOAT

The use of the "Penguin" AGM-119B from SEAHAWK light helicopters gives it the capability to reach naval units located some 18 miles away, adding additional capacity to transport other arms for subsequent attacks.

At the same time that helicopters were being incorporated with a new rotor and modifications to the retractable undercarriage, 116 SH-3H's were being introduced, modified with new radar, sonar and dipole scatterers to confuse older systems. Later it was decided to eliminate some equipment to be able to install new tactical navigation systems and to give it the sonar buoy processor capability.

As a result of the 1982 budget the manufactuer of eighteen SH-60B Seahawk LAMPS MkIII units was begun, a concept which had been validated since 1979 with five prototypes. The entry into service of this advanced platform allowed some SH-3H's to be retired, with the former being introduced to aircraft carriers from 1991.

The close protection Battle Group SH-60F model, with an SH-60R version from 1998, was developed with improved protection to equip modern destroyers. Also ready from 1998 is the Super Seasprite SH-2G optimized with the most advanced cockpit which includes Litton modern display screens, front facing infra-red trackers, equipped with the capability to use anti-boat missiles such as the Kongsberg Penguin AGM-119 B.

The users

Many countries have followed the example of the United States Navy, which has been equipped with early models in their modernized versions, with improved capabilities.

TECHNICAL CHARACTERISTICS

	SH-3D "SEA KING"	SH-2F "SEASPRITE"	SH-60B "SEAHAWK"
COST IN MILLIONS OF DOLLARS:	14	12	20,25
DIMENSIONS:			
Length	22.21 m	16.03 m	19.76 m
Height	5.13 m	4.72 m	5.18 m
Main rotor diameter	18.90 m	13.41 m	16.36 m
Main rotor turning area	280.48 m²	141.26 m²	210.14 m²
WEIGHT:			
Empty	5,447 kg	3,193 kg	6,191 kg
Maximum	9,752 kg	6,123 kg	9,926 kg
Maximum arms loading	1,300 kg	500 kg	500 kg
Fuel	3,714 l	1,779 l	2,233 l
Propulsion plant power	3,320 CV	2,700 CV	3,380 CV
PERFORMANCE:			
Ceiling service height	4,410 m	6,860 m	5,790 m
Maximum speed	272 km/h	230 km/h	272 km/h
Range	1,482 km	661 km	600 km

The SEASPRITE continues flying in the United States and Argentine Naval Air Reserve. The "Super Seasprite" has been purchased by Egypt, New Zealand and Australia. Australia, like Spain, Japan, Greece, Thailand, and Taiwan operates different versions of the SH-60, with two hundred in the navy. With respect to the SH-3, these are in service in Argentina, Australia, Brazil, Germany, Belgium, India, Japan, Pakistan, Peru, and Spain, among others.

Different capabilities

Designed in accordance with different requirements, in distinct periods and incorporating very different electronics and detection systems, the capabilities of the various models analysed differ greatly.

Advanced models

Without doubt the most advanced and expensive of them all is the SEAHAWK which employs a well proven and successful concept, including two powerful navy General Electric jet engines, type T700-GE-410, a RAST recovery system which facilitates landings on escort ship platforms and employ very advanced avionics. Their different configurations include surface search radar, sonar equipment including sonar buoy launchers, magnetic anomaly detectors, acoustic processors, etc. All of which can easily detect any threat, transmitting the information to the carrier ship via a coded communication link, and allowing it to disable the target with its own weapons, or with those of the associated ships.

Continuing with this theme the SUPER SEASPRITE has recently been modernized at remarkably low cost. It now offers a very satisfactory capability with its new equipment and is able to carry out multiple naval hunt missions. The different SEA KING models take a back-seat role despite being modernized. A fact which results from their design and size. Although some units are being directed to complementary activities such as carrying aerial detection systems or the transport of commandos on infiltration missions.

High Powered

The capability of the on board equip-

> **CAPABILITY**
>
> Multipurpose and powerful, the SH-3 heavy helicopter is carrying out service on different naval missions. It is a very effective machine in the search for submarines, a task which it carries out in some of the countries which are using it.

ment is such that it allows the detection of submerged targets or others on the sea surface. But it can do more than detect these targets because the helicopter also has the capability of attacking them its own weapons, these being fixed on laeral mounts on the fuselage and which are specifically designed to carry a whole array of weapon systems. Submerged targets can be neutralized by using Mk-46-5 and Mk-50 light torpedoes which also include a guidance system capable of taking them up to the objective. The aircraft also carriers depth charges which are designed to explode on contact or alternatively at a pre-determined depth. PENGUIN and AGM-114 HELLFIRE Anti-ship Missiles can be used to attack surface targets, these being guided towards the target by using the helicopter's radar tracking system. The 70 mm rockets carried are also very effective particularly when they are fired in salvos against unprotected ship structures. The use of machine gun fire is also an extremely useful weapon, being able to damage very important auxiliary equipment such as radar electronic systems. All of these variations give this helicopter a

weapons system which is able to offer a wealth of different options depending on the country using them, the conditions, and the mission assigned at that time.

It was for features such as these that the SH-60 was used extensively during the Gulf War. Its function was to act as a surface exploration platform being specially equipped with a coded video link which allowed it to transmit real time information captured by its infra-red cameras to its own naval units. An attack was then launched against the target missions in which light machine guns were used as a self-defence measure when faced with possible interceptions by small patrol boats and ships.

The experience of the Vietnam conflict consolidated the use of helicopters for support work during all phases of combat, this included material transportation as well as support for troops on the ground. As in previous experiences, the introduction of the Black Hawk family as a utilitarian machine adapted to modern requirements was the driving force for a wide range of different versions, created to satisfy the most varied of needs.

Requirements

In 1972, after analysing the requirements for the new helicopter established in the study UTTAS (Utility Tactical Aircraft System), the United States Army requested companies to present their proposals to develop the aircraft needed.

Decision

Boeing Vertol and Sikorsky Aircraft were

PROVEN

With the years gone by since its entry into service, the number of units constructed and the conflicts in which it has been used, the design formula of the BLACK HAWK has been validated. It has demonstrated that it was able to carry out all foreseeable missions and that it has great capability to grow with new possibilities of use.

NUMEROUS

Close to a thousand and a half BLACK HAWKS make up the transport helicopter arsenal of the United States Armed Forces to which can be added those units exported to some twenty countries.

MULTI-PURPOSE

Employed in transport or light attack missions, the BLACK HAWK has demonstrated itself to be a multipurpose machine which can be adapted, after the inclusion of the necessary equipment, to carry out all kinds of missions.

---

selected and on the 30th of August 1972 they were each contracted to construct three prototypes which were extensively tested between 1974 and 1975. The YUH-60 flew for the first time on the 17th of October 1974.

It was necessary to construct a fourth

machine for trials, financed by the manufacturing companies, to complete evaluations which lasted up to November 1976. On the 23rd of December 1976 the YUH-60A was declared the winner which was confirmed by the first order for 15 units.

Called the UH-60A Black Hawk in 1979, the first units were sent to the 101st Air Transport Division of Fort Campbell in Kentucky. From this point a production rate of 10 units per month was reached. This has been continued up to the present as there are more than two thousand of these helicopters in service around the world.

Existing Models

The initial version, the UH-60A has been designed to carry three crew members and eleven fully-equipped soldiers. It can be used without any modifications to perform tasks such as medical evacuations, recon-

> **ARMED**
>
> Although it wasn't designed to carry out armed missions the inclusions of two wings allows it to support a variety of weapons from rocket launchers to Hellfire anti-tank missiles, with which it can support ground troops or act as a light attack craft.

is used by the United States Air force to rescue pilots shot down. The SH-60 SEAHAWK is designed for submarine hunting and surface attacks. The AH-60L is used to support direct penetrations with various types of weaponry. The VH-60 is used to transport VIP's (Very Important People) including the President of the United Estates. The UH-60Q DUSTOFF for naval medical evacuation and the CH-60 for general use. This version

naissance, management and control, logistics support, and mine dispersing with the VOLCANO System. This is thanks to some modifications which allow quick and easy changes in configuration. The basic model has been improved since 1989 and its designation has been changed to UH-60L.

The EH-60 is more specialized with the electronic war system QUICK FIX IIB and later modifications, JUH-60A and GUH-60A were used for trials and technical evaluation. The MH-60 is also equipped with weaponry and systems necessary to carry out special operations. The HH/MG-60G PAVE HAWK

> **ADVANCED**
>
> Designed according to some wide reaching requirements the UH-60 has demonstrated itself to be one of the most advanced helicopters of its kind in service at the moment.

has been evaluated by the U.S. Navy since the 6 th. of october 1977. With the result that the company has received firm orders for these machines until 2001.

The users

To those previously mentioned would have to be added the S-70 version for export. Amongst others the Saudi Arabian contract is notable twelve S-70A-1 Desert Hawks with the capacity to transport 15 soldiers with the aid of extra fuel tanks. Sixteen S-70A-IL for medical evacuations with six stretchers, infra-red search lights and improved communication equipment. Two S-70A-5 units employed by the Philippine Airforce. Thirty nine S-70A-9 units assigned to the Australian Airforce, two of which crashed recently when they were transporting SAS commandos. Three S-70A-11 to Jordan. Two S-70A-15 VIP transporters acquired by Brunei. Twelve S-70A-17 used by the Turkish Army and Police, which have been complemented by with fifty more manufactured under licence by TAI. Two S-70A-21 VIP units used in Egypt. Three S-70A-22 VIP used in Korea and others bought by Mexico, The Moroccan Police, Hong Kong and Argentine Armed Forces.

The direct transfers of United States units include the UH-60L's sent to Bahrain, five requested for by Brunei, fourteen UH-60A & L used to combat the Colombian

MEDEVAC

This medical evacuation craft has a cargo hold for four stretchers and a medical team with the possibility of increasing its range of action by optimizing the sub-wing fuel tanks.

narcotics trade. Ten UH-60A YANSHUF assigned to the Israeli 124 Squadron and twelve of the same model lent to the U.S. Drug Enforcement Agency, which monitors the introduction of narcotics and sixteen type L's for Kuwait. As well as these are those units designated S-70A-16, applicable for those produced under licence by the British company GNK Westland. Also to be added are the hundred machines constructed by the Japanese company Mitsubishi since 1988. These correspond to the models SH-60J Jayhawk for search and rescue and the UH-60J for general purpose.

Design characteristics

Widely proven in combat the Black Hawk has demonstrated that its design characteristics allow it to successfully deal with the requirements of a modern battlefield. Noteworthy for its robustness, ease of maintenance and capability to carry out multiple missions.

A resistant machine

During the invasion of the island of Grenada in 1983, some of these machines were attacked by light arms and demonstrated that a fuselage with a combination of tita-

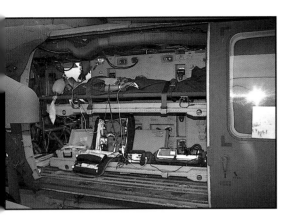

nium, Kevlar, graphite, and plastic fibres was capable of resisting the impacts. It has a fixed landing carriage with pneumatic shock-absorbers incorporated in both front wheels to lessen the blow in the event of heavy falls to the ground. Trials have demonstrated that 85% of the structure remains intact after suffering vertical impacts of 11.5m/s, lateral of 9.1 meters per second and longitudinal of 12.2.

At the same time, projectiles of 23mm diameter can pass through the main and tail rotor blades without causing serious damage; the 1,361 liter fuel tanks are resistant to impacts and include anti-explosion elements; the electrical and hydraulic systems are duplicated; the pilots can count on armored plating protectors on the side walls and below their seats and the cargo deck floor has been reinforced to check light arm fire.

Configuration

Since 1989 it has incorporated two General Electric T700-GE-701C Jet Engines which together generate a thrust of 3,600 horsepower and are connected to the titanium main rotor which includes four blades of 8 meters length, complemented by a small tail rotor positioned at an angle to the vertical.

The pilot and co-pilot sit in the control cockpit which has been designed with instrumentation compatible with night vision goggles. They have at their disposition different flight elements including a double hydraulic system for managing the blade angle, the AFCS Hamilton digital auto-pilot, self-stabilization equipment, the Omega AN/ARN-148 Tracor navigator, Motorola AN/LST-5B satellite communications UHF etc.

Defined as a machine designed by the military for the military, this helicopter represents a landmark in United States industry, and has reached a high level of safety and operational capacity in every combat operation in which it has been employed. A notable feature is its ability to fold the wings and tail rotor for transportation in specific types of planes such as the C-130 HERCULES where there is enough space in the cargo hold for one helicopter, or the C-5 GALAXY which can fit up to five.

Capacity

Designed as a multi-purpose helicopter for multiple missions, the UH-60 is basically a light transport machine which, for example, can carry two infantry platoons or a 105mm artillery piece hooked up on a sling underneath it, with the gunners and ammunition inside.

For self-defence it incorporates passive electronic systems and interference flare launchers in the upper part of the fuselage. At the same time there are anchoring points at the side windows from where the third crew member can operate a medium

or heavy machine gun. In addition the more modern models can have side wings added where auxiliary fuel tanks rocket launchers, Minigun multiple barrel machine guns or up to 16 third generation Hellfire anti-tank missiles can also be added. These require a ground laser or light helicopter such as the Kiowa OH-58 with a laser to guide the missile to the target.

SUPPORT

The third crew member has the responsibility for operating the gun mounting installed in the windows behind the cockpit which is used with medium or heavy machine guns such as the 7.62x51mm SACO M60 or 12.7x99mm BROWNING M-2B HB respectively. Low recoil cannons have also been experimented with.

ROTOR

The main rotor is basically made of titanium and has the job of moving four large blades which give the helicopter the stability and manoeuvrability necessary for its operational missions.

COCKPIT

The pilot and co-pilot cockpit includes all the necessary elements to allow the helicopter to fly night and day in any atmospheric conditions.

UH-60L TECHNICAL CHARACTERISTICS

COST:	5,87 million dollars	Internal fuel load	1,361 l
DIMENSIONS:		Maximum fuel load	6,507 l
Fuselage length	15.26 m	**PROPULSION:**	Two General Electric T700-GE-701C jet engines with a thrust of 1,800 horsepower each.
Height	5.13 m		
Length with rotor turning	19.76 m	**PERFORMANCE:**	
Cockpit volume	11.61 m³	Service ceiling height	5,837 m
Rotor turning area	8.68 m²	Maximum speed	361 km/h
WEIGHT:		Cruising speed	294 km/h
Empty	5,224 kg	Stationary maximum height	2,895 m
Maximum	11,113 kg	Range	584 km
Internal max load	1,197 kg	Extended range	2,222 km
External max load	3,629 kg		

STUB WINGS

Defined as an external load hook-up system, modern models include the possibility of connecting auxiliary fuel tanks or a wide range of weapon systems.

PROPULSION

Accessed by two large sliding doors, the two General Electric T700-GE-701C Jet Engines are located above the transport hold and give the necessary agility for missions entrusted to this helicopter. They are fitted with large outlet nozzles which have special units installed to reduce the infra-red emissions.

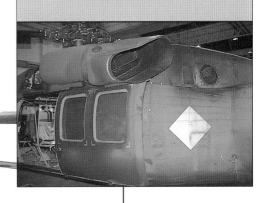

FLARE LAUNCHERS

Incorporated on the fuselage sides are M-130 interference flare launchers which serve the purpose of confusing infra-red seeking missiles and also those guided by radar.

CARRIAGE

The front landing carriage incorporates a pneumatic shock absorption system to which has been added a cable cutting system, and which allows the absorption of impacts up to 11.5 meters per second to reduce the effect of the personnel transporter making an emergency landing.

FUSELAGE

The back part of the fuselage incorporates a small warning light for position, the tactical communication system aerials and a small wheel designed to be able to cope with the vigorous landings which are possible when used in combat.

Widely used to transport troops and equipment during the Vietnam War, as well as limited support and attack missions. The different versions of the UH-1 demonstrated their suitability to meet the requirements that a truly lightweight multi-purpose helicopter must be capable of. Although many of them were lost in combat, their mission capabili-

MANOEUVRES

The Spanish Army's Airborne Forces (FAMET) include various battalions of helicopters equipped with the UH-1H which has demonstrated its robustness and effectiveness to meet any kind of transport or support mission working with ground forces.

MULTI-PURPOSE

The design and features of the IRO-QUOIS has meant that it has established itself in the world at large as a multi-purpose machine which is notable for its low purchase and operational cost.

ties, ease of maintenance, reduced purchase price and mulit-purpose operational and capability characteristics have had a bearing on the fact that they can still be found in service in a variety of armies around the world where it is seen as the ideal solution to carry out the most varied of missions.

Conception

The requirements of the United States Army was to have a multi-purpose helicopter which could serve to evacuate the injured from the front line, an experience greatly felt during the Korean War with the transportation from mobile MASH units to hospitals. This led the BELL Company to come up with a proposal, known as MODEL 204 which ended up as the army's choice during the course of 1955.

Development

After the first flight of the XH-40 which took place on the 22nd of October in 1956. A rapid period of trials began which led to

the order to construct half a dozen YH-40's with the objective of an even greater and ambitious program of appraisal.

The following nine UH-1 helicopters incorporated changes to the original design such as a jet engine with 860 horsepower and canvas seats to transport six people.

Popularly known as HUEY, a name which the manufacturer engraved on the craft's command controls, the supply of the first UH-1A's began on the 30th of June 1959, and were completed in March 1961. later they received the nicknames, SLICK for the troop

carrier, HOG for the gunship version and IROQUOIS for the multi-purpose H model. The A model machines were the first to be sent to Vietnam as a part of the Tactical Transportation Company which arrived at Saigon airport in Tan Son Nhut at the beginning of 1963 to facilitate troop deployment.

Evolution

The UH-1B version began being delivered in 1961 and included a more powerful engine of 960 horsepower and greater load and personnel capacity. From this point many others followed , more optimized and with improved features such as the 767 examples of the C model which had a greater fuel capacity, wider blades and a tail rotor with an inverted

profile. The 2,201 machines of the D model which were propelled by a 1,100 horsepower jet engine and with a capacity to take 12 soldiers, 360 units of which were constructed under licence by the German company Dornier. The E model with modifications to allow it to operate from the amphibious assault craft; the F model supplied with a 1,272 horsepower engine and used by the United States Airforce to support the siting of the Minuteman and Titan ballistic missile silos of the Strategic Air Command. The F model with 5,345 machines manufactured for the United States Army, with the LYCOMING T53-L-13 engine of 1,400 horsepower and elongated fuselage to facilitate the movement of greater loads and personnel.

A great diffusion

This last version is the most popular one which has survived up to the present day with many modifications to the instrument panels, to allow the use of night goggles, for example, and with the installment of new equipment such as radio altimeters, radar emission detectors and supplementary armor for the pilot seats. The D and H series have been given the reference of BELL 205.

The various models have been identified by a base letter preceding the designation. As such, the letter T is assigned for missions of engagement, E for electronic warfare, H for rescue, R for reconnaissance etc. Twin engined versions have also been manufactured such as the UH-1N and AB-212, depending on whether the construction was carried out in the United States or in Italy, and whether it is four-bladed or not as in the BELL 412.

In service

Manufactured in various configurations in factories in the United States, Italy, Germany, Japan, Indonesia and Canada, the different versions of the ubiquitous UH-1 have exceeded 10,000 units, amongst which are 100 BN models (Blade November) requested by the United States Marine Corps from Bell Helicopter Texetron, and to be ready by the end of the year 2003.

In service with the armed forces of more than seventy countries, amongst those are: The United States, Saudi Arabia, Argentina, Bahrein, Brunei, The Czech Republic, Ecuador, Slovenia, Finland, Greece, Guatemala, The Netherlands, Honduras, Israel, Italy, Japan, Morocco, Mexico, Thailand, Turkey, Uruguay and Venezuela. In Spain there are some seventy examples of the H and 212 versions, carrying out active service in the army and navy and they make

ACTIVE SERVICE

Despite the introduction of the more modern UH-60 a total of 1,300 craft continue forming a part of the helicopter wing of the United States Army, which continues to obtain notable results which are derived from its design, mission possibilities, performance and low maintenance work.

EMBARKATION

With twin jet engines and up to date equipment, the hundred UH-1N helicopters used by the marine corps have been modernized to prolong their useful lives.

SUPPORT

To give support fire for its own troops the «H» series can be equipped with multiple machine guns, rocket-launchers, grenade-launchers etc. These give the helicopter an additional combat capacity for all types of missions and support tasks.

up the main element of the Helicopter Battalion (BHELMA) and the Third Air Squadron respectively.

The design philosophy

The BELL 204/205 family of helicopters are designed with a broad cross-section hull, constructed in light alloy and, with the exception of the 412, incorporate a semi-rigid rotor consisting of two articulated blades which determine the plane of rotation by means of a stabilizer.

An innovative helicopter

Huey was the first helicopter of the series which used a jet engine, a LYCOMING T53, installed above the fuselage and close to the main rotor unit, just behind the gearbox unit. This allowed a larger cargo hold and ability to transport more. The

UH-1H TECHNICAL CHARACTERISTICS

COST:	5 million dollars		Internal fuel	945 l
DIMENSIONS:			**PROPULSION:**	A Lycoming T53-L-13 jet engine with a thrust of 1,400 horsepower
Length	13,59 m			
Height	4,41 m		**PERFORMANCE:**	
Rotor diameter	14,63 m		Service ceiling height	3.840 m
Main rotor turning area	168,1 m²		Maximum height	240 km/h
WEIGHT:			Range flying at ground level	512 km
Empty	2.363 kg		Range with auxilliary tanks and	
Maximum	4.309 kg		flying at 1,120m altitude	800 km
Max external load	2.000 kg			

design maintained in its different versions the use of the most powerful and reliable twin motors for flying with the least risk over areas like the sea.

The pilot and co-pilot are together in the cockpit and have access to the interior through side doors, enjoying good visibility to the outside world thanks to large glass windows. The passenger or transport area is accessed by the large sliding doors on the

PILOTS

Sat in seats protected against impacts and provided with helmets which facilitate communications, the IROQUOIS pilot and co-pilot of BHELMA II have the responsibility to operate this aircraft when making night flights for commando infiltration missions.

TWIN-MOTOR

The UH-1N has been given a double turbine which increases its survival possibilities in case of the failure of one of the engines, a basic requirement when it is being operated continually above the sea surface.

side of the craft. This area is where the flight mechanic works and carries out auxiliary tasks. The design is notable for its tail beam inclined upwards with rectangular stabilizers in the central part and a small rotor at the end. It has two skids which make it possible to land in any of terrain.

The normal capacity of the H model makes it possible to transport a dozen fully- equipped soldiers in reconfigurable canvas seats, or depending on need, there could be stretchers or two tons cargo in the hold. Using a sling system it can also carry light vehicles or medium caliber artillery pieces which are hung from a hook inside the helicopter.

In combat

This helicopter was extensively used in the Vietnam War from 1963 onwards and there are many photographs and images

which were widely seen in war films being shown around the world. These images are an accurate representation of the huge of variety of missions performed by this aircraft. These covered activities such as medical air evacuations: the trasportation and collection of troops at different enemy infiltration positions; air patrols of river routes; reconnaisance work; the support of artillery fire and assaults; inteligence work; psychological warfare; giving support fire for ground troops; the combined use of medium, heavy and multi-barrelled machine guns; rocket and grenade launching etc.

Subsequently and in addition to various peace-keeping missions with the UN, models of this helicopter have carried out operations extensively in conflicts such as the Arab-Israeli War, the invasion of Grenada, the peacekeeping operation in Iraqi Kurdistan, the former Yugoslavia, and in operation Desert Storm. During which marine helicopters were used to designate targets at night, thanks to systems such as Nite Eagle which, has an infra-red tracker and laser direction signaler.

Weaponry

The UH-1's weapons can be used with great versatility, such as machine guns used with the cargo hold's access doors open, like the modern MK-2 Marte anti-ship mis-

MODERNIZATION

The BELL 412 is a modernized version of the original design which includes a double turbine, structural improvements and up-dated cockpit and is outstanding as a support element in transport missions.

siles which are carried by the Italian **Griffon** helicopters, a design version which has evolved from the Bell 412.

Basically the different versions of these helicopters can have a variety of weapons fitted in the cargo area as well as externally on the aircraft's fuselage. These are options which depend on the type of missions envisioned for them, but in each case the control of the weapons can be carried out by the operators sitting at the front of the cockpit.

The mounts fitted to the inside of the helicopter are straightforward constructions allowing the operation of both the SACO M-60 medium machine gun and the Browning M-2HB heavy machine gun. There are also automatic 40 mm grenade-launching systems and multiple machine gun assemblies such as the MINIGUN, which has a rate of fire of 6000 rounds per minute.

CAPABILITY

Equipped with two 12.7x99mm heavy machine guns, grenade launchers and rocket launchers the UH-1H is prepared to give punctual and powerful replies to attack actions.

HEAVY DUTY

With its capacity and features the CH-53E SUPER STALLION can be considered capable of transporting more men and cargo than any other machine in service in the Western armed forces. It has also demonstrated itself as a great multipurpose helicopter in those countries which have equipped themselves with it.

Created to satisfy heavy transport needs, the CH-53 has consolidated itself as the helicopter with the greatest capacity for carrying cargo designed and, in service in the West at the moment. Its features and possibilities have been brought out in various battle missions, from operation CSAR to rescue Captain O'Grady, a United States F-16 pilot shot down over Yugoslavia during the UN peace operations to transport civilians evacuated from countries caught up in internal conflicts. These missions have demonstrated its potential for many types of missions.

OPERATION

Incorporated in Marine Expeditionary Units (MEU's) the CH-53E heavy helicopter's main objective is to supply heavy transport vehicles to the marine forces. In addition it transports teams of men and various kinds of material and equipment.

ded to configure a vehicle with the capacity to transport a cargo of 3,630 kg over a distance of 111 miles at a speed of 167 mph. On the 24th of September 1962, after analyzing the proposals of Boeing, Kaman and Sikorsky, the Department of Defence made their decision known. They had decided to develop the SIKORSKY design, awarding the company 10 million dollars to construct a full-sized model, a fuselage for static trials and two prototypes for evaluation.

The first of the YCH-53A flew on the 14th of October 1964, and the evaluations continued without problems. The marines

Development of the CH-53

The period of the SIKORSKY S-65, manufactured as the CH-53, was begun in October 1960 to satisfy a requirements of

the United States Marine Corps which was interested in a heavy helicopter vehicle to replace the Sikorsky HR2S-1/CH-37 Mojave, a model still propelled by radial engines.

Program

On the 7th of March 1962 the BuWeps (Bureau of Naval Weapons) published its requirements for program HH(X) (Helicopter, Heavy, eXperimental) which inten-

received the first batch of 141 CH-53A's in September 1965. Manufactured in the Connecticut factory, the production line continued with 20 CH-53C's for the United States Air Force, 126 CH-53D's for the marines and two CH-53's for the German Army which assembled twenty units in its own factories which were supplied in kit form, to which another 90 were added, manufactured under licence.

Adapted Versions

Other re-powered versions followed from the initial SEA STALLION units, modified for specific requirements, which included two VH-53D's assigned to carry the President of the United States, 47 HH-53H and MH-53J PAVE LOW units, used by the Air Force to carry out rescue missions for pilots shot down over enemy territory, for which they have received very complex equipment and weaponry. 34 MH-53E Sea Dragons were modified by the United States Marines to work as a towing vehicle using a system designed to neutralize naval mines, of these ten are in service with the Japanese Marines designated S-80M, with six more ending up in Iran. A hundred CH-53E Super Stallions, propelled by three jet engines are used in amphibious assault roles by 12 marine squadrons, bringing the total number produced up to 670 units, including two S-65Os, used during the seventies by the Austrian Air Force.

Added improvements

The Israeli firm MATA, of the Israel Aircraft Industries group, has been working since 1991 on the building of 40 units for the Israeli Airforce, initially A,C & D versions of the S-65, to a standard known as CH-53-2000. With a cost of 8 million dollars for each modified unit the modernized helicopters include a fixed nozzle for

DETAILS

A meteorological radar at its bow, fuel tanks mounted on the side of the fuselage, landing carriage consisting of three units with two wheels each, engines located in lateral gondolas, cockpit with good visibility, etc. These are the details which define the United States heavy helicopter.

COMPACT

Once the main rotor is folded up, the size of the SEA STALLION is noticeably reduced and is therefore easily fit into a hanger or amphibious vessel.

refueling in flight, two external tanks on fixed supports to hold 4,000 an additional liters of fuel, a self-defence system with interference flare-launchers, and electronic transmitters. Modifications to the cockpit instrumentation include multifunctional consoles and changes to allow the use of night vision goggles. There has also been a HOCAS system installed to control communication, a seat with controls for a flight engineer has also been included. With these alterations the maximum weight allowance has now increased to 22,680kg.

In combat

It has carried out infiltration missions with special troops during the Vietnam War, transported troops in operation Desert Storm, rescued Apollo crews from the sea, and captured equipment from Israeli commandos who had appropriated an entire SA-6 (Gainful) anti-aircraft missile battery. Evacuated residents in Liberia, Grenada and Albania. The Stallion family of helicopters has participated in two missions which are particularly noteworthy. The first, called Eagle Claw, took place on the 24th of April 1980 with the participation of eight RH-53D's of the United States Marines who tried to free a group of hostages held in Iran. The operation failed because two of them broke down in flight, and a third collided with a HERCULES assigned to do the refueling in flight. However, the second was carried out successfully by a CH-53E on the 2nd of June 1995, the operation was to recover Captain O'Grady who had been ejected over Banja Luka, in Serbian held territory, after his

F-16 was shot down during and aerial mission which was part of Operation Deny Flight.

Great capacity

The potential of the latest CH-53E ver-

move up to 16 tons hung from a sling which is hooked onto the underside of the fuselage.

Its design

The fuselage has a conventional semi-monohull made of aluminium, steel and tita-

CH-53E SUPER STALLION TECHNICAL CHARACTERISTICS

COST:	24.36 million dollars		Maximum load	16,330 kg
DIMENSIONS:			Internal fuel	3,849 l
Height	30.19 m		External fuel	4,921 l
Length	8.97 m		PROPULSION:	
Breadth	8.66 m			Three General Electric T64-GE-416 jet engines
Turning surface area of main rotor	455.38 m²			with a joint maximum thrust of 13,140 horsepower
Turning surface area of tail rotor	29.19 m²		PERFORMANCE:	
WEIGHT:			Service ceiling height	5,640 m
Empty	15,072 kg		Maximum speed	315 km/h
Maximum	33,450 kg		Range	2,075 km

sions is such that it is capable of carrying up to fifty soldiers with basic equipment, or it can

nium, the cockpit has been manufactured with various combinations of glass fibres, other parts have been re-inforced with polyamide panels and titanium. It is a large craft and easily accessed thanks to the large rear door. This is operated hydraulically, which allows the easy entry and exit of men and material. This machine stands out for the internal space available, and its capacity to carry many types of cargo. And the placement of the main retractable landing gear in a lateral gondola equipped with shock-absorbers with a high absorption capacity.

The large tail at the back of the helicopter is slightly angled to the left, with an enormous stabilizer on its upper part. The tail section can be folded back to facilitate the positioning of the machine in transport vessels. In the lower part of the helicopter there are fenders which are automatically extended when the landing carriage is activated on landing. On the front right side there is an access door to the interior which in its upper part includes a winch to hoist small cargo or people during rescue missions.

Propulsion

Three General Electric T64-GE-416 or 419 jet engines are fitted to the machine. A large filter screen is incorporated into each engine casing. Together the engines are capable of producing a maximum thrust of 13,140 horsepower, with a transmission system capable of dealing with a take off loading of 13,500 horsepower. Two of these are located in lateral gondolas in the upper part of the fuselage and the third is incorporated in the structure above the engine, transmitting power to the main seven-bladed rotor and to the four-bladed tail rotor. The blade

TAIL
The tail which includes a four-bladed rotor and large stabilizer can be folded thanks to an automatic system incorporated in the helicopter.

PROPULSION
Between the cargo loading crane and the main rotor are the lateral gondolas fitted with enormous filters at the front to avoid the ingestion of all kinds of particles which two of the three jet engines draw in. The General Electric T64-GE-416 turbines constitute the propulsion plant of the CH-53E SUPER STALLION heavy helicopter.

folding system is automatically activated to reduce its size to the minimum.

To facilitate its maintenance various access and inspection panels have been incorporated into the structure it can also count on an electronic system which gives breakdown warnings and signals the need to replace various components. All of which improves the reliability of a machine which has demonstrated high operational readiness and a reduced number of accidents.

Equipment

Depending on the mission to be carried out, the different members of the CH-53 family can be equipped with a wide range of systems adapted for a particular mission. Usually the different transport variants turn to light and medium machine guns to give cover to personnel and material deployed. Those involved in crew rescue missions in hostile territory include; MINIGUN multi-barrel machine guns capable of a rate of fire of 6000 rounds per minute, and other equipment such as radar and thermal imaging cameras with which they can fly at low altitudes to avoid being detected by enemy anti-aircraft systems; anti-mine units use a non-magnetic vehicle towed along the surface of the water. They are also equipped with a Westinghouse AN/AQS-14 sonar and usually fly equipped with interference flares and electronic disruption equipment for self-defence, using them when necessary. They can also be re-inforced with two types of guided missile: the STINGER or SIDEWINDER.

Widely used as assault craft from the amphibious boats of the United States Navy. The CH-46 has demonstrated for more than thirty years of service its capacity to perform all types of transport activities, especially those related with troops and materials of the United States Marine Corps.

Vietnam, Lebanon, Grenada, Kuwait, Somalia, Liberia, and Albania, among others, are some of the areas of the world where they have been deployed to carry out combat actions, evaluation missions or support tasks for multinational deployments.

Origins

In 1956, the firm Vertol began design and engineering studies in order to develop a twin-jet engine transport helicopter that might have military and commercial applications, and which could use the small and light jet engines which were then starting to be available in the aeronautic market.

Development

After validating the concept in the prototype model 107, which flew for the first time on the 22nd of April 1958, orders arrived for three evaluation models of a military version known as CH-46A, for the United States Army. They would later abandon its incorporation in favor of the larger and more powerful CH-47 Chinook. After

IN COMBAT
Deployed within the SFOR unit sent by NATO to the former Yugoslavia, the CH-46E have performed continuous logistical transport missions as much from their mother ships as from final bases.

nine machines had been purchased by New York Airways, the representatives of the United States Marine Corps saw in this machine the answer to their needs for a large capacity advanced twin-engine model for troop transport during amphibious assault operations.

Acquisition

In February 1961, fourteen CH-46A machines were ordered by the marines. The first of which flew on the 16th of October 1962. Later US Navy evaluations followed. They aquired fifty UH-46s during 1964 for logistical transport operations. In 1965, the Japanese firm Kawasaki received the authorization to construct the helicopter under license and at the same time the United States Defence Ministry ordered the doubling of the production rate in order to satisfy the needs of the Vietnam conflict. Because of this, at the start of the seventies, 164 model A, and 266 model D machines had been built.

MAINTENANCE
The years have passed since its entry into service, and its continual use on all types of missions has required a constant maintenance process in order to maintain the machines at the highest operational level.

To the former followed 186 model F units delivered from 1968 on with additional electronic equipment. There were also some E models manufactured from 1977. They were based on previous versions in which more powerful engines were installed and modifications to the pilot's seats were made in order to reduce the impact in case of falls to the ground. Also, the HH-46D and RH-46 were employed respectively on rescue and mine-hunting missions on behalf of the Navy; the KV-107II/IIA was constructed under licence in Japan, of which 17 were exported to Saudi Arabia. To these we must include the CH-113 "Labrador", acquired by

the Canadian Air Force to carry out SAR missions, the CH-113A "Voyager" which were delivered to the Canadian Army between 1964 and 1965, and the 55 Kkp 4C constructed for the Swedish Navy and Air Force. Of these last units, 45 were assigned to submarine and mine-hunting operations.

Improvements

Called "Frog" by its crew, almost 300 CH-46E Sea Knights are currently still in service. The majority of them are used by the Marines in 16 active and 2 reserve squadrons. To these can be added the twenty machines used for crew training. Their current operational availability is due to the various improvement processes that have been applied to the original machines, such as the CILOP carried out during the seventies in order to replace the rotor blades with lighter and stronger fiber-glass ones. Also, a more powerful propulsion unit was added and structural reinforcements made that increased its survival capabilities in case of

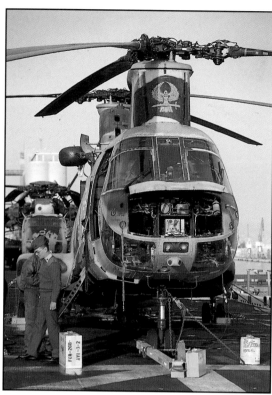

capacity substantially increased.

Additionally, defense systems have been installed such as flare launchers and electronic countermeasure equipment. Paint with a better infrared signature has replaced the original, and some areas have been reinforced with armor plating. The cockpit now includes warning radar connected to an infra-red generator to confuse first generation ground-air missiles. The instrument panels have now been adapted for the use of night goggles and some machines have been equipped with the Communications Navigation Cockpit System (CNCS) which includes digital radios, navigation equipment and a global positioning system (GPS). For its part, the Navy is introducing to its machines a Dynamic Component Upgrade (DUC), which has been contracted with Boeing Defense and Space since 1995.

accident or during combat missions. Meanwhile, at the end of the eighties, the Service Life Extension Program (SLEP) and Helicopter Emergency Flotation System (HEFS) meant improvements were introduced, with the latter to increase its emergency landing resistance capability, having had the fuel tank

Design

Compact and capable, the CH-46 suffers from some limiting features inherent to the age of its design. Although throughout its years of service, it has shown its durability and efficiency to fulfil its assigned activities.

Structure

This is the result of a fuselage manufactured in aluminum alloy, designed in order to achieve the greatest interior space and access for cargo and personnel. In its upper part are the reinforcements that support the transmission and the propulsion unit. Entry is through a rear access ramp and in the interior situated on the sides, there are two rows of seating. There is capacity for up to twenty-six fully-equipped soldiers, or to transport all types of combat support equipment, from light missile-launchers to munitions and logistical support. It can also be adapted for medical evacuations with fifteen stretchers and two medics. A hook situated on the undersideallows a crane to winch up cargoes of 4,535 kilograms.

Propulsion

Two powerful and reliable General Electric T58-GE-16 Jet Engines, improvements on the original version, each rendering a maximum of 1,870 horsepower, give it the necessary power to complete the assigned missions, and be able to return to base with only one engine working.

These engines turn two large three-blade rotors in tandem, one in the front part and the other in the rear section. They turn in opposite directions to improve stability during stationary flight. Their blades can be retracted rapidly through a system controlled from the cockpit. This greatly facilitates the storage capacity on deck or in the hangars

> **ROBUST**
>
> After thirty years unbroken service, the "Sea Knight" design has shown itself to be very robust and adaptable to the basic needs of capacity and weight required by the United States Marines, which plans to replace the current fleet before the year 2010.

> **COCKPIT**
>
> Provided with a large glass windscreen which favors visibility, and equipped with radar emission detectors on the outside. The cockpit has been configured with simple and operational analogue equipment.

of the amphibious ships that normally transport them. This is also useful when maintenance tasks need to be carried out.

Tactics

During the Vietnam era, when it flew at high altitude to avoid being hit by light anti-aircraft fire. The machine nowadays carries out missions such as Terrain Flight Missions (TERF) in which the machine flies at an altitude of between 15 and 20 meters or less, in order to take advantage of the ruggedness of the terrain and to make it difficult to locate. Assaults *en masse* combine the advantage of surprise with the difficulty of enemy reaction against multiple targets. The capabilities of the crew are put to the test in The Nap of the Earth Flight (NOE), in which it flies at an altitude of less than 15 meters along previously studied routes. They are also challenged in the Combined Arms exercises (CAX) which are carried out in the Californian desert zone of Twenty-nine Palms. In both operations, on board personnel are in charge of operating through two side windows two 12.70 x 99mm Browning M-2 heavy machine guns that constitute its defencive and support armaments.

STERN

On each of the sides of the rear section, a half-gondola is located housing the rear undercarriage components, different types of electronic equipment including anti-skid panels on the upper part to facilitate maintenance tasks in this area of the fuselage.

PROPULSION

Incorporated in the upper rear part of the fuselage, just below the tail rotor, are the two General Electric T58-GE-16 Jet Engines propulsion units that produce a total thrust of 3,470HP with which the agility and capacity of this medium-type helicopter is guaranteed.

ARMED

A small window on each side allows the siting of a medium and heavy machine gun for self-defense tasks, easily carried out when the window is removed. For this there is a mounting permitting exact lateral and vertical movements of the weapon.

DETAIL

The complexity of the cabling and fluid pipes integrated into the fuselage are obvious in this shot taken from the right-hand side door, which includes an access ladder to facilitate the entry and exit of the crew.

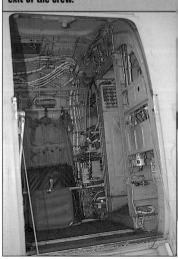

ROTORS

The rotors, which include a triple-bladed head, can be retracted automatically and correspond to a very effective although outdated design. They are ideal for transport missions that involve low and medium altitude flight.

EQUIPMENT

The forward hold, in the helicopter nose, houses a large part of the electronic equipment and systems associated with the cockpit instrumentation and contains the components necessary to ensure safe flight over the sea and in any weather conditions.

FRONT UNDERCARRIAGE

Robust, with pneumatic shock absorbers to reduce the impact on the two steerable wheels, the front undercarriage is characterized by its simplicity and small size.

TECHNICAL CHARACTERISTICS CH-46E

COST:	16 million dollars	Maximum internal cargo	4,000 kg
DIMENSIONS:		Internal fuel	3,786 l
Length	13.66 m	**ROPULSION:**	
Height	5.09 m	Two General Electric T58-GE-16 jet engines	
Rotor diameters	15.24 m	units producing a total thrust of 3,740 HP	
Rotor surface area	364.6 m²	**FEATURES:**	
Hold area	16.72 m²	Service ceiling	4,265 m
WEIGHT:		Maximum velocity	267 km/h
Empty	5,927 kg	Range	383 km
Maximum	10,433 kg		

sales of the AS-330 PUMA helicopter, with 700 units produced and flying in many countries, it was decided to go ahead with the construction of an advanced version with a variety of improvements to the original model with the idea that it was to be more of a military than a civilian helicopter.

Evolution

The first preparatory studies by the French for a new version of this much appreciated transport helicopter began in 1974 and continued until a prototype was ready for trails, which first took place on the 13th of September 1978. After complying with a series of verification trials, the production phase for the AS-332 SUPER PUMA began halfway through 1981.

In July 1983 the L version with an elongated fuselage was certified and in 1986 the propulsion plant was modified with the introduction of the Turbomeca Makila IAI engines. The military versions produced from 1990 were given the designation AS-532 and called the Cougar to differentiate them from those machines destined for civilian operators.

Sales

The machines leaving the production line of the French company Aerospatiale, or those produced or assembled under licence in Indonesia, Spain, Turkey, or Switzerland are identified by a letter code which defines its characteristics and mode of use. As such U corresponds to unarmed units, A for armed, C for combat with a short fuselage, L for a long fuselage, and S for antisubmarine or anti-ship naval missions. Since 1992 the Mk2 has been sharing the production line with the older Mk1.

Nearly 500 units of different versions of this family have been ordered by close to seventy users in 67 countries, two thirds of which are for military use and 28 for VIP's (Very Important People). These include the UB/AB which has only been on offer for a short time, responding to the needs of the most modest of budgets.

T he French defence industry knows how to optimize its aeronautical developments to satisfy the multiple requirements demanded by a variety of countries. Demonstrating with the Super Puma that it can design and manufacture transport helicopters with advanced capabilities to support ground, naval, or air operations. This has led to it having an impressive order book which it is predicted will grow with the new variations expected to be introduced in the next few years.

The conception of the helicopter

After the success encountered with

> **ADVANCED**
>
> The latest version created for military use is the AS 532 US which includes an elongated cabin with space for up to 29 soldiers; navigation systems, side winch, modified engine outlets etc which allow the most varied of missions to be carried out.

Military users

Among the most notable of the users are the French, who have specialized units such as the Horizon battlefield radar detector and the Mk2 U2 assigned to rescue work, the CSAR (Combat Search and Rescue), The Spanish who have ordered more than forty units for the Air Force and Ground Army, Abu Dhabi which uses five modified to have sonar and the capacity of launching EXOCET AM-39 anti-ship missiles and torpedoes, Turkey which operates twenty which it assembled under licence and there is a list of

Conception

It has a large semi-monohull fuselage and is manufactured using a light aluminium alloy which acts as a base to be reinforced with titanium and composite materials. The front part is assigned for the flying of the craft, which requires one pilot in good visibility and two in times of IFR flight. In the center is the cargo hold and depending on the version, there is space for up to 25 soldiers and their personal equipment. In the back is the tail and rotor, on the sides the Messier-Bugatti landing carriage designed with high-

other countries which, amongst others, include Slovakia, Venezuela, Zaire, Oman, Brazil, Chile, The Netherlands, China, Ecuador, Panama, Mexico, Nepal, South Korea, Japan, and Germany which has assigned them to its border police.

The design

It is praised for its capacity to maintain operational with 98% effectiveness in support work ca-rried out on the North Sea oil platforms, yet criticized by others, such as pilots in Spain's Airborne Forces (FAMET) who don't consider it to be very satisfactory in its daily support role. The truth is that this French model is becoming more and more successful.

absorption parts to cope with possible ground impacts.

Five fuel tanks are positioned along the fuselage and under the cabin floor, with a capacity of 1,497 liters for the UC version and 2,141 for the SC version, it is possible, to install flexible cabin tanks to carry an additional 2000 liters additional, and with 2 exterior tanks, each holding 325 liters, it is capable of carrying out long range missions.

Propulsion

Located in the upper part of the fuselage and occupying a large part of its length, we find the propulsion unit with the two Turbomeca Makila IAI Jet Engines at the

to improve their resistance to impact and perforation. While the tail blades combine carbon fibres, resins and titanium.

There are two independent systems for operating the moving parts, and to connect them to the different flight control elements. Electrical energy is supplied by two 20kVA 400Hz alternators.

front. The design is a modular one with each turbine having a maximum power of 1,877 horsepower, integrating air intakes with filters in front to prevent the ingestion of ice and foreign objects. Some models have been modified to reduce the infra-red signature. Mk2 models are supplied with the Makila IA2, with 2,109 horsepower.

The engines turn the transmission system up to 23,840 revolutions per minute and transmit their power to the main rotor which turns at 265 r.p.m. and to the tail rotor at 1,278 r.p.m. The four main blades are completely articulated and have been manufactured with elastomers

Great capability

It was originally designed to transport troops with combat equipment, which it can do using individual seats or with sol-

iers seated on the fuselage floor. Different
versions are offered which support its
capacity to receive different weapon systems or to move weights up to 4.5 tons
hung from a sling. Machine guns can be set
up on the mounts inside the aircraft for its
self-defence, and on the outside walls it is
possible to mount 20mm GIAT monotube
cannons, 68mm rocket-launchers, light torpedoes, Exocet or Sea Skua anti-ship
missiles etc.

Amongst some of the more noteworthy models are the VIP version, with
an optimized cabin to transport important people in the greatest comfort
possible, battlefield surveillance units
with the Horizon antenna system located
in the lower back part of the fuselage, and
naval support units with a variety of sonar
equipment including sonar buoy launchers, and display consoles. The rescue
CSAR, used in France and Saudi Arabia
incorporates a fixed nozzle for in-flight
refueling, and a side winch for hoisting
loads up to 272 kg, optics for night time
flights and search missions, auxiliary floats
built into the lateral landing carriages,
weapon mounts, and a cockpit with digital
displays and the most modern equipment,
making it possible to rescue two pilots in
a zone located 500 miles from the depar-
ture point and to be able to return to
base without refueling in flight, something

SELF-PROTECTION

Peace missions, as executed by this machine assigned to the United Nations, require the use of flare-launchers to interfere with the seeker head of anti-aircraft missiles (photograph above).

SPECIALIZATION

Two COUGAR AS 532 UL units are assigned to transport the Horizon battlefield detection system which can simultaneously process the information coming from 4,000 targets (photograph to the right).

which, if done, would considerably increase this distance.

TRANSPORT

Assuming the most varied of missions the French COUGAR helicopters have been specifically designed to transport troops and support elements in every kind of military operation (photograph on the left).

AS 532SC TECHNICAL CHARACTERISTICS

COST:	14 million dollars			Maximum external load	4,500 kg
DIMENSIONS:				Internal fuel	2,141 l
Fuselage length	15.53 m		**PROPULSION:**		
Height	4.92 m			Two Turbomeca Makila 1A1 jet engines with	
Main rotor diameter	15.60 m			a combined power of 3,754 horsepower	
Main rotor turning area	191.13 m²		**PERFORMANCE:**		
Cockpit surface area	7.8 m²			Service ceiling height	4,100 m
WEIGHT:				Stationary ceiling height	2,800 m
Empty	4,500 kg			Maximum speed	278 km/h
Maximum	9,350 kg			Range	870 km

COCKPIT

The most advanced features of the COUGAR cockpit include four large digital presentation screens and numerous indicators associated with the various flight parameters , elements which facilitate the work of the pilots located in this front part of the fuselage.

WEAPONRY

Simple mounts on the inside of the cockpit allow the personnel transporter to employ light and medium machine guns with greater accuracy, although when it is necessary lateral mounts can be adapted for more significant weapons.

CABLE CUTTERS

An extremely sharp blade is incorporated in the front part of the fuselage to cut all kinds of cables, avoiding accidents as a result of low altitude flight.

LANDING CARRIAGE

The front of the retractable londing carriage inclu-des two small wheels, located below the cockpit, while the two main elements are located on the sides and are drawn up into the lateral modules on the fuselage.

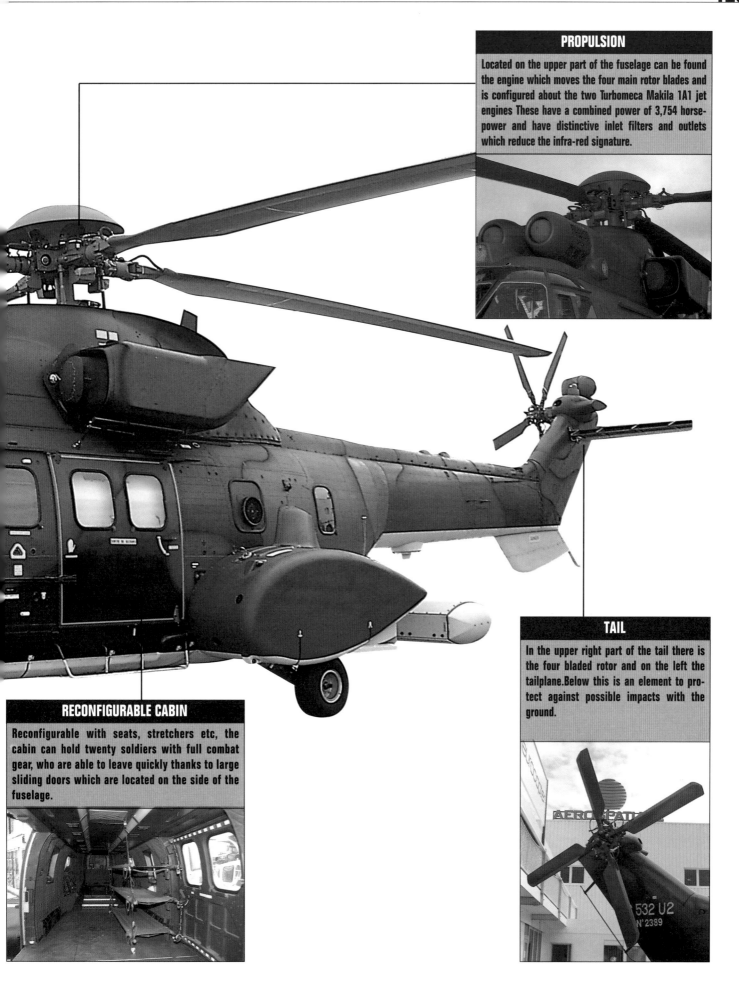

PROPULSION

Located on the upper part of the fuselage can be found the engine which moves the four main rotor blades and is configured about the two Turbomeca Makila 1A1 jet engines These have a combined power of 3,754 horse-power and have distinctive inlet filters and outlets which reduce the infra-red signature.

TAIL

In the upper right part of the tail there is the four bladed rotor and on the left the tailplane.Below this is an element to protect against possible impacts with the ground.

RECONFIGURABLE CABIN

Reconfigurable with seats, stretchers etc, the cabin can hold twenty soldiers with full combat gear, who are able to leave quickly thanks to large sliding doors which are located on the side of the fuselage.

Considered as one of the most powerful of western helicopters, the Chinook, is distinguished from the rest by its capability to transport all types of cargo. Widely tested in combat, from Vietnam to the Falklands, it has demonstrated its robustness and range of use in many situations. The Chinook is able to carry a large number of troops, and carry large items of equipment by using the fixed points under the fuselage.

Development

As a reply to a United States Army

OPERATION

With its features, possibilities and performance, the CH-47D "Chinook" is one of the most capable transport helicopters, for this reason more than a thousand have been manufactured and are now in service in about twenty countries across the world.

OPERATIONAL

Various CH-47D's have been operating within NATO's forces assigned to the transportation of soldiers, weapons, and materials of every kind.

requirement issued in 1957 for the supply of a heavy helicopter VERTOL, which merged with Boeing in 1960, offered its CH-46 design which was considered to be insufficient. Because of this, the bigger V-114 was created with greater endurance designed for the US Army first as the YHC-1B in 1962, and then later as the CH-47 "Chinook".

Evolution

A YHC-1B prototype was revealed for the first time on the 21st of September 1961 and was quickly put through a process of validation along with four others. This culminated in a formal request for the helicopter to go into production. The CH-47 was propelled by LYCOMING T55-L-5 jet engines with 2,200 horsepower each and with a capability of moving 2,790 kilograms over a distance of 111 miles, or 6,068 kg over a distance of 22 miles.

The first units were received on the 16th of August 1962 by the First Airborne Cavalry Division which would use a large part of the 354 units produced of this version.

On the 10th of May 1967 the CH-47B was introduced with more powerful 2,850 horsepower jet engines, of which 108 machines were manufactured. On the 14th of October in the same year the CH47-C flew which led to the delivery of 270 units, each with 3,750 horsepower T55-L-11A engines. In 1973 a modernization process began on 182 CH-47 units with the introduction of new rotor blades manufactured with composite materials. Using faster inspection methods and in 1976, the process of reconfiguring the CH-47D began with the oldest models.

After the flight of the prototype in 1979, a production program was began which resulted in deliveries in February 1984 to the 101 Air Transport Division. After which 470 units of the new CH-47D were requested and delivered up to 1994 to the United States Army, which wants to keep them in active service until 2025, by which time its replacement, currently known as the AAN is due to be in service. Other modernization programs, such as the I.H.C. are also under consideration at present.

Specialisation

Special versions such as the MH-47D used by The Special Operations Command Regiment have come out of the original design. These following on from the AH-47 GUNSHIP version of the Vietnam era. They are capable of being re-supplied in flight and are equipped with thermal imaginy cameras, BENDIX/KING RDR -1300 meteorological radars, two 7.62 MINIGUN mul-

THE UNITED STATES

The United States ground forces operate with around five hundred CHINOOKS to support both normal and special missions with another two hundred being used by the National Guard.

SPAIN

The airforce units of the Spanish Army (FAMET) employ seventeen CH-47 units in the Helicopter Transport Battalion (BHELTRA) around the Madrid area.

tiple barrel machine gun assemblies, radar warners, electronic war equipment, ELBIT ANVIS-7 night vision goggles and a hoist winch with a capacity of 272kg. The machine also has the capability of transporting 44 soldiers. In addition there are the GCH-47 models which have been configured for engineering training work and the Mk2 of the British Royal Air Force updated to a Mk3 version for the special forces.

Exports

Of the thousands of units manufactured by Boeing; 10 have gone to civilian companies, 5 to Argentina where three operate in the army and two with the airforce, 12 to Australia, 9 to Canada, 2 to Japan, 6 to the Netherlands, 6 to Singapore, 24 to South Korea, 19 to Spain which lost two in accidents and where the rest are assigned to the Helicopter Transport Battalion (BHELTRA), 3 to Taiwan, 6 to Thailand, 58 to Great Britain, and 734 to the United States of which 540 have been modernized.

A production license has been given to the Japanese company KAWASAKI which is manufacturing 56 machines, of which 40 are for the country's army and 16 for the airforce. The Italian company AGUSTA has produced 134 machines, destined for Egypt, Greece, Iran, Italy, Libya, Morocco and the Pennsylvanian National Air Guard which has received 11 units manufactured in Italy.

Characteristics

The design which led to its considerable size and cargo capacity are characteristics very much appreciated by the those in the military who work with this model.

Fuselage

It has a constant metallic cross-section, very wide and elongated, with five small observation windows along the side. If also has a large door at the back which allows the rapid disembarkation of troops or material, even when hovering over uneven terrain. A side door on the front right part of the fuselage and an advanced cockpit with lateral windows allow the pilots to escape in times of emergency.

The need to carry out flights in any weather condition has meant that the latest model includes every kind of advanced display to help the pilots including: radar altimeter, automatic stabilization, VOR receiver, TACAN, complex HF & UHF communication equipment, horizontal position indicator, satellite links, global positioners etc. The cockpit has been modified to allow it to carry out night flights using night vision goggles without any compatibility problems with the instruments. For daytime flights visibility is very good with the large front window and with two additional crew members who have responsibility for observation at the large rear door and side windows to facilitate the work of the pilots.

The fuselage sides are prominent with the front section containing the electronic systems and the rest incorporating fuel tanks, which have a capacity of 3,899 liters, and associated equipment. On the underside

TECHNICAL CHARACTERISTICS

COST:	30 million dollars		Maximum load	12,284 kg
DIMENSION:			**Internal fuel**	3,902 l
Length	30.14 m		**PROPULSION:**	
Height	5.78 m		Two Allied Signal T-55-L-714 turbines with a joint thrust	
Rotor diameter	18.29 m		of 8,336 horsepower	
Rotor turning area	525.3 m²		**PERFORMANCE:**	
WEIGHT:			Service ceiling height	3,095 m
Empty	10,693 kg		Maximum speed	298 km/h
Maximum	24,494 kg		Range of action	185 km

there are the four fixed landing carriage units, the front two with two wheels each and pneumatic shock-absorbers, the back two with one wheel each which can be orientated.

Propulsion

Two Allied Signal T55-L-712 SBB jet engines are located externally in two lateral gondolas on the large tail rotor structure, each with a maximum of 4,314 horsepower. The 714 model has an increased power of 4,867 horsepower in times of emergency. Two very large rotors, each made up of three blades, turn at 225 revolutions per minute, coupled to a gearbox which can cope with power loadings of up to 7,500 horsepower. These blades are constructed in such a way that they have a honeycombed nucleus, covered with glass fibre laminate which can resist impacts of up to 20mm. The rear rotor is positioned significantly higher than the front one.

The engines, which have proven themselves to be very robust and which have a low incidence of break-downs, are fed by fixed fuel tanks on the fuselage and by other auxiliary tanks with a capacity of

CAPABILITY

The ramp at the rear allows fast embarking and disembarking of troops and material, while the hook-up fixings below the fuselage permit the transportation of all kinds of cargo like these two 105mm Oto Malera light artillery guns with their ammunition and equipment.

COCKPIT

The pilot and co-pilot enjoy a cockpit designed in a rational manner, equipped with all the necessary instrumentation to facilitate their flight and transport operations.

9,084 litres located in the cargo hold. The APU Solar T62-T-2B auxiliary power unit drives a 20kVA generator and hydraulic pumping system which produces the necessary electrical and hydraulic power, independent of external support.

Capacity

The large size of the hold, with a surface area of 21 square meters and volume of 41 cubic meters, permits it to be configured for a variety of different kinds of personnel transport thanks to the seats which are strategically incorporated along the sides in a manner allowing space for 44

SPACIOUSNESS
The size of the fuselage and the rear ramp gives this machine a sense of spaciousness which markedly differentiates it from other similar designs.

the 155mm M198 gun which is hung from a sling, while the gunners and associated equipment including ammunition are transported inside the helicopter.

Self-defence is trusted to a radar emission detector, interference flare-launchers and when necessary two 12.7mm Browning M-2HB medium machine guns can be installed at the door and window behind the cockpit. These are operated by the two additional crew members, who make up a total crew on board of four people. The British models have TRACOR chaff-launchers, HONEYWELL AN/AAR-47 alarm systems for the approach of missiles, LORAL AN/ALQ-157 and GEC-MARCONI AR118228 RWR infra-red interference systems. This is equipment derived from its extensive use in real combat missions.

fully-equipped soldiers, this can be increased to 55 in times of emergency.

The large size of the rear door allows the embarkation of light vehicles, pneuma-

CAPACITY
On the underside of the fuselage the four fixed wheels are noticeable, making up the landing carriage. Also clearly seen are the fixing points which allow it to move cargo by using slings (photograph to the left).

tic equipment, artillery guns, missile systems etc. It also allows for the fast disembarkation of parachutists in either the traditional manual or automatic way. In the bottom part of the fuselage are hooks for hoisting cargo, the central one with a capacity of 11,793 kg and the other two 7,711 kg. The combination of these allow the quick movement of heavy artillery pieces such as

FUSELAGE
With its compact, advantageous design, the CHINOOK's fuselage stands out from other designs because of its spaciousness and possibilities for transporting all kinds of cargo (photograph on the right).

Following the opening up process of other ex-Soviet companies, KAMOV has been incorporated into the military and industrial company Mapo. It has recently been promoting two new attack helicopters which comply with the requirements demanded of aircraft which have to confront the threats of the 21st century.

The first known as Ka -50 Hokum-A, is a single-seater attack machine characterised by its compact size and some notable features. While the second, the Ka-52 Hokum-B, is an older, re-equipped single-seater version. To help with their exportation, they have been christened the Black Shark and Alligator respectively.

EVALUATION

With the fuselage painted in desert tones more in step with use in the Middle East, the various BLACK SHARK prototypes are being consciously evaluated with a view to their possible adoption as a standard attack helicopter in a variety of countries.

CAPABILITY

With its design, special features, and multi-purpose weaponry, the Ka-50 stands out from other models currently being offered in the world market.

The origins

Helicopters like the Ka-27, Ka-29, Ka31RLD and Ka-32 took up a large part of the purchases of the Soviet Navy. After which the company KAMOV worked on the design of an attack machine which would satisfy the requirements of the Russian Army and which was conceptually similar to the HAVOC with which it competed inside and outside the Russian market.

Conception

After initial studies the design of the Ka-50 project was ready at the end of 1977 and work began on the manufacture of the the first prototype, initially known as V-80Sh 1 (VERTOLYET-80). This flew for the first time on the 27th of July 1982, its existence became known in the summer of 1984, when the first photographs of it were published in the 1989 report "Soviet Military Power", published by the United States Defense Department. Western specialists were able to observe and get to know it during the 1992 Farnborough Air Festival where it was called the Werewolf, a name substituted later on by the existing one.

Development

The objective was to have 12 pre-production

units manufactured at the Arsenyev factory, two of which were moved in 1995 to the Russian Army's aviation training center in Torzhok. During which time work continued on the verification and implementation of this single-seater attack machine. Known as the Alligator, with the code name V-80ash2, it was presented at Le Bourget Festival in 1995 and flew in 1996. Incorporating some advanced electronic equipment from France in those units destined for export. It can be deployed as a transitional, operational, or training machine for pilots heading for combat units.

Evaluations

Its adoption has been considered by some countries including Slovakia which negotiated the purchase of six single-seaters to equip an air force squadron. The objective of Sergei Mikheyev, a doctor of

technical sciences and chief designer of the company Kamov, is that some of his two attack helicopters will be adopted by the Russian Armed Forces.

Capability

Although the Russian models are manifestly technically inferior to their western equivalents, their robustness, reliability, low purchase price, low running cost and capability as multi-purpose machines make them attractive in a variety of markets.

The design

Following the line in the positioning and type of propulsion used by previous models of this design company, the new KAMOVS are characterized by being the first single-seater helicopters in the world specializing in attack missions. With a thinner fuselage in the single-seater and more rounded one in the twin-seater these models are highlighted by the positioning of a large part of the flight sensors and target detectors in the front nose and a ball with additional sensors positioned above the cockpit for future versions.

Heavily armored- it is estimated that 750 kg of material has been used- the structure has been constructed around a torsion box principle which includes various apertures with access panels for internal equipment

and as such improves access for maintenance work. It stands out for its weaponry, retractable landing carriage, and it is provided with low pressure wheels to allow it to operate on any type of terrain. There are two large stabilizers located in the upper part of the helicopter and in the tail there is a directional rudder instead of a rotor.

Propulsion

Two KLIMOV jet engines TV3-117KV with a thrust of 2,190 horsepower each drive the single-seater and an earlier version has been modified to increase the power up to 2,465 horsepower to drive the two-seater craft. They are positioned on the upper parts of the sides of the fuse-

The pilot or pilots, depending on the model, are located in a strongly protected cockpit provided with armored glass which protects them from direct impacts of up to 23mm fired from 100m, and at the same time includes a unique evacuation system, called the ZVEZDA K-37, which consists of an activator which explodes the rotors and fires a rocket to sweep the crew towards the outside where they can then use their parachutes to land safe and sound. The associated flight equipment includes ground tracking radar, HUD data screen similar to the MIG-29, VHF, UHF and HF communication systems, friend-enemy detector, ground map displays etc.

lage to prevent anti-aircraft fire from hitting both of them at the same time. The entrance nozzles have spherical deflectors positioned in front of them to make it difficult for foreign bodies to enter and the

TECHNICAL CHARACTERISTICS

	Ka-50	Ka-52	PROPULSION:		
COST IN MILLIONS OF DOLLARS:	Unknown	Unknown		2 KLIMOV V3-T117VK Jet engines with a total thurust of 4,380 horsepower	2 KLIMOV V3-T117VMA Jet engines with a total thurust of 4,930 horsepower
DIMENSIONS:					
Length	16 m	13.53 m (fuselage)			
Height	4,93 m	4.95 m			
Rotor diameter	14,5 m	14.5 m	PERFORMANCE:		
Surface area each rotor	165,13 m²	165.13 m²	Stationary ceiling 4.000 m	3.600 m	
WEIGHT:			Maximum speed 350 km/h	300 km/h	
Operational	9.800 kg	10,400 kg	Ascent capability	600 m/s	480 m/s
Maximum external load	2.000 kg	2,000 kg	Combat operational range	250 km	200 km
External fuel load	2.000 l	2,000 l	Design factor	+ 3 g's	+ 3 g's

exhaust nozzles are designed to offer the lowest possible infra-red signature.

Power is transmitted through a gearbox to the counter rotation double rotor system, which has three blades each. Each of these rotors has been been designed to resist light impacts and can be dismantled for transportation in the hold of cargo aircraft like the 11-76. This kind of configuration makes a tail rotor unnecessary and dispenses with the sophisticated and vulnerable transmission systems that run through the upper part of the fuselage to transmit power from the propulsion unit. A small jet engine can be found on the upper central part of the fuselage which acts as an auxiliary power source to allow the starting up of the KLAMOV and which also supplies electric and hydraulic power for support work on the ground.

In combat

Identical weapon systems can be found in both of the models, these being associated with the active and passive avionics sys-

tems incorporated on board the aircraft. These give the helicopters the capability of detecting and finding their objectives over very long distances, during the day or night, or even in the most adverse of weather conditions.

The helicopter's powerful 30 mm 2A42 cannon assembly can be found on the left side of the fuselage. This assembly is maneuverable because of a hydraulic system which allows the cannon's position to be changed by up to 30 degrees vertically and then 6 degrees laterally, and which is fed by two magazines located in the center of the fuselage. These magazines have the capacity to hold 460 rounds of either armor-piercing or high-powered explosive shells. The stub wings are positioned on the sides of the fuselage in such a way that they almost appear to be truly functioning wings, but which instead allow the loading

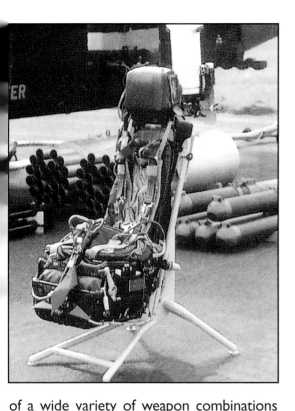

of a wide variety of weapon combinations able to reach a total of some two tones in weight. Launchers are positioned at the extremities of the stub wings with a capacity for holding up to 128 flares. These perfom the function of a self-defence system, acting as decoys against enemy missiles tracking the helicopter and its thermal signature.

There are also two further positions where the following variety of equipment can be installed: AT-X-16 Vikhr anti-tank modules with the capability of holding up to 6 laser guided missiles with a range of some 10 kilometers; non-guided B8V20A rocket-launchers which can hold up to twenty S-8 rockets; the B-13 rocket laun-cher with a total of five S-13 rockets; the UPK-23-250 housings which are used to hold two 23 mm cannons; 250 and 500 kilogram free-fall bombs; KMGU submuni-tions launchers; large 500 litre fuel tanks and Igla-V air-to-air missiles. It has the capability of carrying and then launching Hegler AS-12 anti-radiation missiles and Aphid AA-8 and Archer AA-11 air-to-air missiles. This capability has been highligh-ted because it gives this helicopter the possibility of engaging a variety of diffe-rent kinds of enemy fighter bombers and attack aircraft in combat.

The need of European armies and their associated industries to carry out a series of significant actions created the concept of the Eurocopter Attack Helicopter-Tiger, which is presented as an advanced response within the growing field of specialized machines created to meet the current demand.

Versatile, capable, powerful and, above all, European. This helicopter has been developed thanks to the combined efforts of France and Germany, by reaching self-sufficient production neither ruled out that other countries like Sweden, Spain, and Australia (were the PT4 prototype crashed during trials in February 1998) will end up going for this model to replace the existing air-fleet dedicated to attack and anti-tank missions.

LAUNCHES

Acceptance trials have allowed the integration of the ATAM Matra system in the "Tiger" so that it can deploy MISTRAL air to air missiles to attack other helicopters or aeroplanes in flight at high or low altitude, and its possibilities for combat missions are significantly increasing.

CAPABILITY

We can see a STINGER air to air missile being fired by one of the prototypes configured in accordance with the German request for a multipurpose attack helicopter capable of different types of missions.

Conception

The planned program to substitute, during the 1990's, the attack helicopters the "Gazelle" and Bo-105, in service in France and Germany respectively, brought the governments of both countries to an agreement to develop an anti-tank helicopter. The first contacts were made in 1984 and on the 13th of November 1987 a firm decision was made; the development contract was signed on the 30th of November 1989 by the Eurocopter company formed by the French Aerospatiale and German Messerschmit Bölkow Blohm (MBB) companies.

The Evaluations

The initial forecast was for the construction of three unarmed prototypes for aerodynamic evaluation. One configured for tank-hunting, designated "Gerfaut", and another completed to carry out mixtures of attack and reconnaissance missions, reaching a climax with the flight of the PTI on the 27th of April 1991, from which construction continued until the completion of five helicopters on the 21st of February 1997.

As well as static and vibration trials, inspection of fuselages, verification of equipment integration etc, diverse flight trials have been executed which include the launching of HOT-2 missiles by the PT-5 in May 1997 up to the validation of its ability to

operate in Arctic conditions, for which the PT4 carried out exercises in Sweden equipped with skids below the wheels to land and take off in snow covered land at -30°C.

Supplies

Although the constant reductions in the defence budgets of European countries participating in the program have influenced the quantity of equipment produced, and the period foreseen for entering service. The signing up for production phase, on the 30th of June 1995, gave impetus for the programmed introduction into service in the first few years of the 21st century.

In addition the PT4 has already fired its

PROTOTYPES

Five prototype units are being used to validate the features of the "Tiger", which it is hoped will go into production in the first few years of the 21st century.

Deployment possibilities

Although the "Tiger" is immersed in the development and pre-production phase, it is demonstrating that its deployment possibilities and mission capabilities are as were hoped. It is a design which incorporates the latest advanced technology to take it into multiple types of missions.

Advanced characteristics

Conceived using the most advanced technology such as a structure which is made up of 80% composite materials without rivets, titanium, aluminium, Kevlar etc which gives it the capacity to withstand collisions with the ground of up to 10.5

front cannon and launched MISTRAL air-to-air missiles and non-guided rockets, in response to the HAP French support and escort validation requirements. Meanwhile, the PT5 has launched the STINGER air-to-air missile, corresponding with the UHT/HAC multi-purpose combat and support version. The French foresee incorporating the 115 HAP and 100 HAC, while the Germans have asked for the 212 UHT. A delivery process will begin in 2001 that could last up to halfway through the second decade of the 21st century, that is, if the estimated number of units to be supplied is not reduced.

meters per second, and equipped with two MTU/Rolls Royce/Turbomeca MTR390 jet engines which each produce 1,285 horsepower and 1,558 in times of emergency. With exhaust nozzles integrated in the fuselage to reduce its signature, the TIGER is an attack machine born to neutralize present and future threats in every type of environment and situation. The four-bladed main rotor is constructed combining elastomeric fibres and composite materials which have an indefinite life-span, the landing gear is fixed and permits landings of up to 6 meters per second, and there are redundant hydraulic, electrical and fuel supply systems

The cockpit

The cockpit has been configured to minimize the crew's work. Manufactured in a tandem design with the pilot positioned at the front and the gunner in the back, the cockpit incorporates the latest avionic advances with multifunctional color presentation screens, control display units (CDU), automatic flight control systems (AFCS), intercommunication systems (ICS), radio frequency indicators (RFI), alarm display screen, and weapon control panels. The pilot and gunner can use helmets with protection visors onto which can be projected a variety of flight data and parameters and which also have an integrated holographic system to facilitate weapon-firing. This has elements to facilitate communication between the different data parts or with systems which are on the same side. Self defence is assigned to the Thomson -CSF TSC2000 system with an IFF friend -enemy interrogator, communication and laser-threat alarm, and interference systems launcher.

Capability

The TIGER has been configured as a multipurpose helicopter, based on a modular concept, with the possibility of being equipped with various systems which allow it maximum operational flexibility and to be able to assume new roles whenever necessary; as well as adapting itself to the continuous changes in international politics.

The HAP variant includes television and infra-red sensors, FLIR (Forward Looking Infrared), for nocturnal flights, measuring distances by lasers with optical elements to capture and follow targets. These are installed in a giro-stabilized housing SFIM/TRT STRIX positioned behind the upper part of the pilot's cockpit. Its weaponry consists of an automatic cannon GIAT M-78 positioned below the nose, provided with up to 450 thirty millimeter rounds, while on the wings there are two fixing points where 12 or 24 rockets (68mm SNEB) can be positioned. The machine is also equipped with armor piercing darts and four MISTRAL air-to-air-missiles.

The UHT/HAC includes a mast above the rotor which supports the OSIRIS Euromep Sensor conceived from a second generation targeting system which includes FLIR, laser and TV, while below the nose a FLIR is integrated for the purpose of flying.

Its weaponry consists of rocket-launchers, heavy machine gun batteries or light

HAP

The French version HAP is a new generation combat helicopter with air-to-air and ground support capacity, for which it has a 30mm cannon in the turret, 68mm rocket launchers and MATRA MISTRAL air-to-air missiles which make it a difficult adversary in combat.

TIGER

This French-German combat helicopter can carry out its missions at night, in bad weather and is capable of carrying a wide range of weaponry which enable it to manage a variety of threats and defensive missions.

cannons. STINGER air-to-air missiles will be fitted to the machines ordered by Germany and anti-tank missiles in the HOT-2 and TRIGAT versions.

Up to four of these can be positioned on each wing, the first guided by the SACLOS system and the second being of the fire and forget type. The HOT-2 has been validated in trials carried out between May and June in 1997, in one of which shooting was done with the helicopter flying backwards at a speed of 140 km/hour, while to homogenize the TRIGAT it is fired at by a PANTHER helicopter, with modifications incorporated in the TIGER from 1998.

INDEX

F-18 «Hornet» 4

Rafale 10

F-15 «Eagle» 16

"Gripen" 21

Eurofighter 2000 26

F-14 «Tomcat» 32

Tornado 37

The Harrier 42

Mig - 29 «Fulcrum» 48

Mirage 2000 54

The latest generation Sukhoi 59

F-16 «Fighting Falcon» 64

Air to air missiles 70

European naval helicopters 75

AH-1 «Cobra» 80

AH-64 «Apache» 86

MIl MI-28 «Havoc» 92

United States naval helicopters 98

UH-60 «Black Hawk» 102

UH-1. A highly developed family 108

CH-53 E «Super Stallion» 113

CH-46 «Sea Knight» 118

From the Super Puma to the Cougar 124

CH-47 «Chinook» 130

The attack Kamov 135

«Tiger» 140

Index 144